THE PERCEPTION OF COLOR

THE PERCEPTION OF COLOR

RALPH M. EVANS

A Wiley-Interscience Publication

JOHN WILEY & SONS, New York · London · Sydney · Toronto

Library of Congress Cataloging in Publication Data:

Evans, Ralph Merrill.
The perception of color.

"A Wiley-Interscience publication."
Bibliography: p.
1. Color-sense. 2. Perception. 3. Color—Psychology. I. Title. [DNLM: 1. Color Perception. WW150 E92p]

QP483.E9 152.1'45 74-10812
ISBN 0-471-24785-5

Printed in the United States of America

10 9 8 7 6 5 4 3 2 1

PREFACE

Scientific method and modern instrumentation were first applied to color about 150 years ago, and within the present century the measurement and specification of the stimuli that produce color have developed into the sophisticated science known as colorimetry. We can now not only measure such stimuli with a precision approaching the discrimination sensitivity of the eye but also, for a somewhat restricted range of applications, calculate how to produce stimuli which exactly match each other.

Considering the rather advanced state of this art, it is rather surprising to find that there has been relatively little advance in the subject as far as it concerns what an observer actually sees from a stimulus or group of stimuli, for which the specifications are known. This is particularly true of what would appear to be the simple case of the difference between two colors which are nearly alike. For this case, however, we do have approximation equations, and much work is being devoted to their improvement. For the general case of two or more quite different colors next to each other we still have little more than broad generalities.

Somehow the obvious complexity of this phase of the subject, the many theories of color vision, the conflicting suggestions as to how perceived colors can best be arranged, and the failure to find a simple and accurate "uniform chromaticity diagram," have led to the notion that perceived color is not at present derivable from knowledge of the stimulus. I suspect that most workers, if pressed, would say that this phase is necessarily awaiting the discovery of a wholly acceptable theory of color vision and cannot be solved without it. While this may be true for the ultimate details of the bewildering array of visual phenomena, the very existence of the many widely divergent theories already available strongly suggests the possibility that too much emphasis has been placed on theorizing and too little on studies of what is to be explained.

In any case, the discovery in 1958 of a new visual threshold, by myself and my associates S. M. Newhall and R. W. Burnham, and the subsequent studies of this new threshold with B. K. Swenholt have led to a complete reexamination of the variables of perceived color in relation to their psychophysical analogues. This in turn has led to an attempt to reorganize the subject of color perception in accordance with the new facts, without regard to theories of color vision. The result of that attempt is the subject matter of the present book. In writing it I have been confronted with the task of presenting the new organization required and also, at the same time, of presenting the new facts themselves. This has led to a necessarily diffuse treatment which at times may confuse the various arguments. This is, I am afraid, inevitable in any radically new treatment of a subject based on new facts.

There has emerged from this work what I feel is clear evidence that a satisfactory solution exists to the problem of calculating the colors a standard observer would see from given stimuli under given conditions, without the necessity of introducing any hypothetical mechanisms. The text thus envisions the possibility of extending the range of what I call "stimulus colorimetry" into the new field of "perceptual colorimetry." It requires, however, the addition in all cases of at least one dimension to the three known sufficient for the specification of a stimulus and, for the general case, of two dimensions. It also must make allowance for the fact that, in the general case, the colors seen may be varied over a specifiable range by a voluntary change in the observer's attitude. These variations come about from the demonstration that there are five distinct and independently variable color perceptions, rather than the three usually assumed.

I have often found it necessary, in carrying through the logical consequences of some phases of the subject, to flatly contradict some very firmly held beliefs. In many cases I have been just as reluctant to give up these beliefs as will be any of my experienced readers; yet I am convinced that to a newcomer the new statements will be far more acceptable than the ambiguities which we have so long accepted. In fact it is the resolution of just these ambiguities and inconsistencies that gives me faith in the fundamental soundness of the new approach.

The present text has not been written hastily; it has been three years in preparation, and my questioning and probing of the subject go back over at least twenty years. In fact it was dissatisfaction with the perceptual parts of the subject presented in my *Introduction to Color* according to the then-accepted views that prompted my investigation of the concept of saturation and which resulted in the discovery of the new threshold. The fact that this in turn has resolved the saturation problem to my satisfaction gives me considerable gratification in spite of the resulting upheaval of the subject.

It is unfortunate that neither time nor space is available to elaborate on the consequences of the results presented here. It has been necessary to merely suggest some of them in the last chapter. It is my hope that many of these subjects will be given the required development by other workers.

No attempt has been made to give a complete historical account of the subjects discussed nor complete bibliographic references to the relatively enormous literature of color perception. Where a pertinent concept has been introduced, I have tried to at least lead the reader toward further information on the subject in the many excellent books now available. For this reason I have included in the bibliography a number of books for which there is no reference in the text. Much of the material, however, particularly in Section II, comes from our own work. Most of this has been published in the articles cited.

Readers familiar with the psychological literature may be concerned that I do not make a distinction between perception and sensation. It has been my intention throughout to avoid as far as possible all theorizing as to the mechanism of visual processes. For this reason I have chosen to use the word "perception" in its broadest sense as essentially synonymous with "seeing." It thus includes sensation as a necessary element and also includes conscious mental processes where applicable. The subject matter, however, is limited to color and this term is used in a rather severely restricted sense which will become apparent from the text. The terms "color perception" and "perceived color" are thus not descriptive as to mechanism but are definitely limited as to subject matter included.

One of the major contributions of the text is that the appearance of fluorescent materials finds its logical place in the general subject of perceived color for the first time. The finding that the appearance is not due to the fluorescence is, in fact, one consequence of our discovery of the new visual threshold and is developed *in extenso* for the first time here. Recognition of the fact that the appearance can be produced without fluorescence and that fluorescence does not necessarily produce the appearance is in one sense the key to the resolution of many of the former ambiguities of the subject. It also demonstrates clearly that these ambiguities have been due almost entirely to the historical oversimplification of the subject as far as perception is concerned. In the same connection, gray and the grayness of colors, a subject which has always been relegated to subordinate roles by scientists, is shown for the first time to be the important separate variable that it has always appeared to artists.

The absence of color illustrations in a non-mathematical treatment of color may seem surprising at first. It should be apparent from the text, however, that any such attempt would confuse rather than clarify the development of the subject by placing undue emphasis on a small, although important, phase of the subject. It is partly the assumption that all phases of the subject can be illustrated on the printed (or hand-painted) page that has restricted development of the subject as a whole. One of the challenges the book presents for future teachers of the subject, in addition to the reorganization involved, is the development of equipment for demonstrating all the perceptions and their interrelations. The requirements should be clear from the text and I have often demonstrated all of them to groups of considerable size. It requires, however, that the demonstrator be able to influence the sensitivity level of his observers' eyes, both for intensity and for color.

Perhaps equally surprising is the fact that the important subject of the scaling of the perceptual variables is hardly mentioned. It is a basic tenet of the book that most of the work devoted to such scaling has been based on incorrect, or at least incomplete, assumptions as to the variables to be scaled. My purpose will have been accomplished if I have made clear what these variables are. It would be presumptuous to also attempt to analyze the large literature on the subject in the new terms, even assuming it were possible. The new variables, however, should be easier and less ambiguous to study, at least from the standpoint of the observer.

It would be a pleasure, if it were possible, to list the names of all those to whose concepts I am indebted in arriving at these results. Where I am aware of published material I have referred to it in the text, but many of the ideas must have come from casual discussions and broad reading in the subject and are long since forgotten as sources. I can only thank all those I have known in the field and regret that so many of them are no longer living.

In the course of both the experimental work which led up to the conclusions here stated and in the preparation of the present manuscript, I am particularly indebted to Bonnie Swenholt who, both as co-worker who has shared in the generation of the ideas and as patient critic of the many versions of the text as it evolved, has made many suggestions and prevented many errors. Such errors as remain, however, either of detail or of concept are clearly my own responsibility.

Acknowledgment is due to the *Journal of the Optical Society of America* for permission to reproduce Figures 4-4, 4-5, 4-6, 7-1, 8-1, 8-2, and 9-1.

Ralph M. Evans

Rochester, New York
June 1973

As stated above, the concepts presented are those of Ralph M. Evans. However, since his untimely death on January 29, 1974 prevented his final review of the manuscript, I must accept responsibility for any errors of detail that remain.

Bonnie K. Swenholt

Rochester, New York
March 1974

CONTENTS

SECTION I
SINGLE STIMULI

SECTION II
RELATED STIMULI

SECTION III
THE OBSERVER AND THE GENERAL STIMULUS

THE PERCEPTION OF COLOR

SECTION ONE

SINGLE STIMULI

One *INTRODUCTION*

Color vision and color perception are unique subjects in several respects. First, they deal with one of the major sense receptors of the body, the eye, and its primary stimulus, light. They thus involve the still largely unknown physiological interactions of these receptors with the brain. Second, in spite of the fact that of all the senses vision has received the most attention from both physicists and psychologists over the last 200 years, the results are still very much open to discussion. Third, unlike the physical sciences but in common with other psychological subjects, the end result of a cause cannot be measured but only described, or compared to the effect of a similar cause.

The physical stimuli for vision, as well as for the other receptor senses, such as hearing or touch, can be defined quite rigidly. Color vision differs from the other senses, however, in a relatively simple and well-established characteristic that makes possible the grouping of the infinity of possible stimuli into points in a three-dimensional array. This permits all possible stimuli to be defined, within prescribed limitations of the visual mechanism, by three relatively simple variables. This in turn has made possible the development of the relatively exact science of colorimetry within the subject of color vision.

The techniques and concepts of colorimetry permit us today to describe in unambiguous terms, all possible light stimuli that lead to the perception of color. This simplification of the stimulus phase of color perception now permits, with careful consideration of the assumptions on which colorimetry is necessarily based, a systematic approach to the subject that was not available to earlier workers in the field.

We can thus start with a single stimulus that is the only thing visible to the observer and consider what he sees in terms of all the possible physical variables of such a stimulus. We can then gradually increase the complexity of the viewing situation by the addition of other stimuli in the neighborhood of the stimulus under consideration. If sufficient complexity is introduced this can include all possible viewing situations from the standpoint of the physics of the stimuli, the lights reaching the eye of the observer. It is then necessary to consider the observer as a thinking individual and investigate the extent to which his knowledge of (or assumptions as to) the causes of the various stimuli may modify his color perceptions. In this way we can hope to encompass all possible relations between the stimuli for color as described by colorimetry and the actual colors perceived. It is not yet possible to carry this scheme through to completion; but we can fill in many of the details and at least sketch the probable nature of the final solution.

This approach to the subject is not new; it is essentially that followed by Helmholtz in his attempt to consolidate all the knowledge of his time, along with his own investigations, into a systematic theory of color vision. This great work was hampered, in retrospect, by his somewhat too firm conviction of the uniqueness of the eye response to stimuli and by a theory that, while it explains much, is inadequate for a total explanation. Without either limitation we can now carry matters considerably further. We can do so, however, only because we now know that some perceptions that appear to be due to the stimulus itself are in fact due to the relationship of the stimulus to its perceived environment. Again, this is not a new fact but one now seen to be far more important and far-reaching in color perception than originally thought.

It is interesting and important that throughout the somewhat violent history of our subject there has been little disagreement on the facts involved. The disagreements that have caused the invectives have centered almost wholly on the proper explanation of the facts, each worker feeling that his theory explained everything. As a matter of fact it is largely due to the conviction of the early workers that their theories were correct that we have such a great wealth of descriptions of visual phenomena. Each worker appears to go to great lengths to find facts in support of his theory. This is perhaps an unfair comment on the tremendous scientific interest in the subject in the latter part of the 19th century, but it highlights the fact that essentially all the phenomena of color perception now known were mentioned in some form by at least one writer of the period, usually in support of a theory.

With the perspective of 100 years we can now see that the major differences between these theories were due in large part to the phase of the subject that their proponents considered of paramount importance to their point of departure, so to speak, in investigating the subject. A historical review of the development of the various ideas involved has not been attempted, but it is interesting to note the approaches used by the principal workers simply because, taken together, these form essentially the outline of the present study.

The scientific aspects of color, as we know the subject, had their origin in Newton's demonstration in the late 1600s that sunlight could be analyzed by refraction into invariant color components and that these could be reunited in various ways both to form other colors and to re-form colorless light. Of course, knowledge of color and speculation about it long predated Newton. There was a sufficient body of literature in fact, and this was so persistent, that Goethe, well over 100 years after Newton's demonstration, felt justified in his contemptuous reference to the event as introducing "the epoch of a decomposed ray of light." Nevertheless it was on Newton's findings, amplified by the studies of many workers culminating in Young, Maxwell and Grassmann, that Helmholtz built the great foundations for the science of colorimetry. It is the science that, by unequivocally defining the light stimuli, provides the necessary stepping stone for an eventual science of color perception.

Goethe, working in the late 1700s (his Farbenlehre was published in 1810), represented the pre-Newtonian school of thought and attempted to reconcile it with the scientific attitude of the time. Although an extremely keen observer and a competent experimenter, he sought to conform his thinking by means of his findings rather than approaching the subject with an open mind. Our immediate interest in his work lies mainly in two facts on which he based the whole subject. He starts with the observed phenomenon of the *adaptation of the eye to light.* In the fifth sentence of his work he writes, "The retina, after being acted upon by light or darkness, is found to be in two different states, which are entirely opposed to each other." He thus neatly introduces the basic phenomenon and his main theory in the same sentence, but he amplifies the phenomenon in line with then current thinking in the ninth sentence where he says, "If we pass suddenly from one state to the other, even without supposing these to be extremes, but only, perhaps, a change from bright to dusky, the difference is remarkable, and we find that the effects last for some time." The other fact on which he lays great emphasis is the overriding importance of the perception of white. While we will not accord it the primacy he bestowed on it, it is a fundamental fact easily neglected.

Hering, somewhat later than Helmholtz, took as his starting point the perception of what I shall call "object color." His first sentence, "When we open our eyes in an illuminated room, we see a manifold of spatially extended forms that are differentiated or separated from one another through differences in their colors," sets the stage, and most of his brilliant observations throughout his work are made in this environment. We are, in fact, indebted to him for much of our specific knowledge in this field.

The work of Helmholtz, in connection with the others, established beyond question the fact that all color stimuli can be grouped into families of stimuli (now called metamers) which look alike but differ spectrally, and the fact that each such family of metamers can be specified by three independent variables. Note that both these facts hold true for *all* observers, but the makeup of the families may differ from one individual to another and the number of variables needed for specification may be less (never more) than three.

From this established fact, Helmholtz then proceeded to the logical assumption that the colors seen could also be described by three perceptual variables. To the best of my knowledge, this assumption has been accepted universally by all workers in the field, although a number have pointed out that it is not a necessary conclusion from the facts. Although at first sight Hering proposed six variables, he still believed that any given color perception could be described by a maximum of three because his variables were mutually exclusive in pairs; and he was careful to point this out.

This assumption has, in fact, become such an article of faith with color workers that to question it at all is essentially heresy. All my earlier writings were based on this assumption and my *Introduction to Color* is an explicit attempt to cover the whole subject on this basis. I now feel that much of the awkwardness and distortion of the subject has been due to this assumption and that there are actually at least five perceptual variables involved, reducible to three only under special circumstances. It may be well to point out, however, that I do not question the fact that all *stimuli* can be specified uniquely by three variables. The other variables arise from the context in which a particular stimulus is seen.

Katz, working in the early 1900s, strictly from the standpoint of a perceptual psychologist, introduced many new variables into the subject. These had to do largely with the appearances of the stimulus aside from those due to its spectral energy distribution. As such, they provided a very important and much needed bridge between color perception *per se* and the much broader general field of perception in vision, but they extended the meaning of the word "color" far beyond the limited meaning I shall use. Not all the perceptions due to spectral energy distributions alone can be handled with only three variables; this fact became heavily obscured through the introduction of these added, extraneous variables. From our standpoint, however, Katz made a tremendous contribution to the subject through his emphasis on the perception of illumination as a separate phenomenon caused by the situation as distinct from the localized stimuli, that is, illumination as distinct from objects. This is in fact his point of departure for the whole subject; it immediately eliminates the possibility of a one-to-one relation between the local physical stimuli and the perceptions they produce. He states, for example, in the sixth sentence of his Introduction: "With the same immediacy with which we perceive the colors of objects comes the apprehension of their illumination; and the illumination is not limited to the objects alone, for the empty space between objects is also seen as illuminated." While we need not pursue the latter part of his thought, the first part is essential to an understanding of color perception.

These earlier workers thus represent, through their major points of view, the essential elements of the subject of color perception. Oversimplified: Helmholtz—the physics and psychophysics of the stimulus; Goethe—adaptation and the importance of white; Hering—object colors; Katz—illumination as a separate perception. All of them also presented theories with which, in general, we shall not be concerned except as an aid in presenting the facts.

There has been considerable confusion in the terminology used throughout the subject, largely because of the many points of view introduced. Much of this has been due to the insistence on three perceptual variables, in particular the ambiguous use of "saturation" by Helmholtz. But perhaps more of it has been due to the use of words which imply the situation in which a perception occurs *in addition* to the perception itself. A case in point is the use of the word "brightness" for light sources and "lightness" for object colors with the implication that they are, in fact, the same perception. Thus we read statements to the effect that "in the case of object colors the perception corresponding to brightness is called lightness." We shall find that the transition thus lightly passed over is one of the basic problems in color perception.

Over the years there has been much discussion, some of it rather heated, of the words "perception" and "sensation." While it is undoubtedly desirable to make a distinction between the concepts implied when a theory of color vision is to be discussed, it is not necessary to do so in an overall view of the whole process; one can assume simply that there is, necessarily, an initial physiological response to the light and that this undergoes neurological processing before reaching higher brain centers and consciousness. In this way "to perceive" becomes essentially synonymous with "to see" and the effect can be related directly to the external cause, provided all factors are taken into account. The inclusion among these factors of such matters as the "mental set" of the observer is part of the subject of perception and this still does not preclude relating the effects directly to the presented stimuli, even though their effect may be only to determine a range of possible perceptions (short of hallucinations) rather than a specific single one. In line with this, and part of the problem in the past, is the wholly unwarranted assumption that because a light stimulus can always be defined by a single spectral energy distribution function it therefore, necessarily, produces a single perception. In daily life, the observer often, and perhaps usually, sees more than one color as parts of the same stimulus area; it is a necessary prerequisite for the perception of illumination separately.

It has become customary to divide the subject of color into three broad fields called "physical, psychophysical, and psychological." While the boundaries of these concepts are and have been subject to debate, they are extremely convenient for presentation of the facts if used as broad categories. I shall use the word "psychophysical" to imply measurable characteristics of the eye's color response that have sufficient general validity to be considered properties of the visual mechanism and restrict "psychology" to those cases in which the observer as a conscious individual plays a part in what is seen. Those who wish to identify psychophysics with sensations and include both in psychology will find no deliberate contradictions in what follows.

The word "color" has at one time or another been assigned to one or more of these categories by various writers or groups and in unmodified form has become almost completely ambiguous. It is thus necessary for each writer to now state his intended meaning. In what follows I use the word as though it related exclusively to the psychology category, but in a much narrower sense than is usually used in that subject. As used it is essentially undefinable, having a parallel, perhaps, in the word "taste" as it applies to materials placed in the mouth, and having the antonym "colorless" as roughly parallel to "tasteless." The restrictions will become apparent from the text, and a modifying adjective will be used where it appears needed. One point, however, should be noted about our use of the word: it is not synonymous with the word "appearance"; it represents one small phase of the broad subject of visual appearance, which is itself a small part of the general subject of perception.

One of the outstanding facts about color perception is the large variability between individuals in the stimuli that to them produce matching colors. There has been a tremendous amount of investigation into this variability over the last 100 years and it can well be claimed a separate subject in itself. Somewhat surprisingly, however, it is not necessary to consider this in connection with our study of color perception. However important it may be for any theory of the nature of the visual mechanism, we need only to consider the normal case in order to include all the observed variables. As already noted, no *added* perceptions are found for the abnormal cases. For this reason the discussion will be restricted not only to the normal case, but specifically to the internationally agreed upon CIE Standard Observer (see Chapter 4) to the extent that the specifications cover our needs.

At the risk of redundancy, we might also note here some other restrictions that will be placed on the subject matter; none of them affects the number of perceptual variables involved in our concept of color perception and all of them are discussed briefly in their appropriate contexts.

Perhaps the most unfortunate, but necessary, assumption is that all separate stimuli must be considered as perfectly uniform physically over their individual areas (cone angles at the eye). This is necessary in order to be able to define each stimulus by a single set of numbers. It is doubly unfortunate because few uniform stimuli are ever seen in ordinary surroundings and because the nonuniformity itself produces a large effect on the stimulus-perception relationship.

Two other assumptions are made in the following chapters, largely to keep the work to a reasonable length. It is assumed that neither variation of stimuli with time nor extremes of intensity introduce *new* perceptual variables. Thus I do not discuss the perception of flashes of light or intermittency effects, and do not consider scotopic vision in detail although there is a large literature on both.

It might seem that by thus limiting the subject matter we have too greatly simplified matters. We shall find, however, that they are adequately complex.

Two

PHYSICS AND PHYSIOLOGY

Light, considered as a purely physical phenomenon, is a complex subject. Its various ramifications comprise the science of physical optics for which there is a large literature. From the standpoint of color perception, however, only a small part of this subject is relevant. This smaller part of the subject has also received adequate treatment in recent years, although usually as an introduction to the subject of colorimetry, the measurement of color stimuli. One of the best of these is Judd (1952) and, at a different level, Billmeyer and Saltzman (1966). The present book can be considered a revision and extension of the perceptual phases of the subject as set forth in my earlier book, Evans (1948), which covers in some detail all the physical phases necessary for what I shall consider here.

In the subject of color perception itself we are concerned only with the physics of light as it enters the eye. While the various transformations in quality and direction that the light may have undergone from the time it originated at some source to the time it finally reached the eye are often quite complex, the characteristics of the light at this point can be considered to be very simple indeed. The whole subject of color thus divides sharply into the physical causes that produce the relatively simple final stimulus and the complex physiological and psychological reactions to this stimulus that produce the perception of color. Since these two complexes are bridged by the light entering the eye it is worthwhile, at this point, to consider this light thoroughly and from an elementary point of view.

When any material is heated to a sufficiently high temperature it radiates (gives off) energy into the space around it. This energy, once emitted, travels through space in straight lines at high velocity (3×10^{10} cm/sec). This speed of propagation is so great that for all except astronomical distances its arrival at the eye may be thought of as simultaneous with its departure. When it encounters an object in its path it is either absorbed, transmitted, or reflected. If transmitted or reflected it continues in straight lines as long as the medium in which it is traveling is homogeneous. Its velocity, however, depends on the medium through which it is traveling, always being less than when traveling in a vacuum: the ratio of velocity in space to that in the medium is called the "index of refraction" of the material.

The energy so propagated can be thought of as in the form of electromagnetic waves vibrating at right angles to the direction of motion, and with frequencies characteristic of the originating source. These frequencies are invariant and are thus not affected by the medium in which the wave is traveling. Such radiation is a quite general phenomenon comprising x-rays, radio transmission, and so on, depending on its frequency range. The eye is sensitive to only slightly more than an octave of frequencies in the range of 4 to 8×10^{14} cycles per second (hertz). Light is defined as radiant energy within this range. Ordinary colorless light, such as daylight, can be thought of as comprising a continuous distribution of all possible frequencies, varying only in the amount of energy in the different frequency regions. Because the high frequencies involved in light were found difficult to measure, it has become customary in the subject of color to use the more easily measured wavelength rather than the more logical frequency. This is usually stated with respect to the velocity in a vacuum, and the wavelength range of light is thus usually stated as somewhat exceeding the octave from 380 to 760 nanometers (nm), one nanometer being 10^{-9} meter. It is not of concern in perception that the wavelength of a given radiation is not independent of the medium in which it is traveling, because we deal with it only after it enters the eye, in which it has a constant relation to frequency.

Ordinarily, light can be thought of as vibrating in all planes at right angles to its direction of propagation. Under certain circumstances light may arrive at the eye vibrating primarily in a single plane and is then said to be "plane polarized." While it is possible to see an effect due to such polarization (Haidinger's brushes), there are no known effects which need concern us in the study of color perception.

Energy distribution with wavelength is thus the *only* pertinent physical variable of a steady beam of light except for the direction from which it arrives at the eye. The various directions, in conjunction with the lens of the eye, form an image on the back of the eye (the retina) in which a uniform area in the scene is represented by a uniform area in the image. The light in this image will, of course, have been modified slightly by the eye lens and the aqueous medium behind it and so cannot be considered identical with the light from the scene. Although this light is the true stimulus for color vision, the effects of the eye lens are essentially constant for a given individual and so can be considered as part of his eye sensitivity characteristics. Accordingly, in this book, when I refer to the "stimulus," or the "physical stimulus," for light perception I shall be referring to the light *as it reaches* the eye. For our purposes, this light can be described *completely* by stating the amount of energy it contains as a function of the wavelengths within the range over which the eye is sensitive. It has only the *two* variables, wavelength and energy.

By passing a narrow band of colorless light through an appropriate optical device (prism, diffraction grating, etc.), we can spread out the energies in the light along a wavelength axis so that the energy at each wavelength is available separately. The band of light so formed is called a "spectrum" and may be seen to vary in color from violet, at the short-wavelength end, through blue, green, yellow, orange, and red, the whole simply dying out at both ends as the response of the eye fails.

Such spectra are ordinarily continuous, that is, energy is present at all wavelengths over the regions in which there is any energy. For this reason it is customary to represent the energy-wavelength relation by a curve in which energy is usually the ordinate and wavelength the abscissa. Such a curve is called a "spectroradiometric curve" and completely describes the light. It is often more useful to describe the energy in relative rather than absolute terms. This can be done either with respect to the energy distribution of a known source or with respect to some fixed energy level. In either case the curve is then referred to as a "spectrophotometric curve." A special case of the latter, in which the light is that reflected from or transmitted by an object, is in frequent use. The energy is then expressed relative to 100% reflectance or transmittance and the curve named accordingly.

Spectra are not always continuous; the energy may be concentrated in one or more narrow wavelength regions or there may be one or more narrow regions in which no energy is present. These are called line and absorption spectra, respectively. The former is characteristic of sodium or mercury vapor lamps and the only common case of the latter is sunlight. Both continuous and line spectra may be present simultaneously. This is true of all present day fluorescent lamps and so is of widespread occurrence.

Thus, since energy can be absent or present in any amount at any and all wavelengths, there are an infinity of possible spectral energy distributions. There are, however, certain types of distribution that are sufficiently important to have received special names. Perhaps the most important for present purposes is that in which all the energy is concentrated in a single narrow band of wavelengths. Light that has this distribution, is called, interestingly enough, "monochromatic," meaning "all of one color." It is also often referred to as "spectrally pure" or just "pure" and is defined by its central wavelength. Such light obviously gives a family of unique reference points and we shall have frequent occasion to use the concept.

Another type of energy distribution with unique properties is that which is radiated from a small aperture in an otherwise complete enclosure. This radiation is unique in that its complete spectral energy distribution depends only on the temperature of the cavity and can be calculated if this temperature is known. It is called "blackbody" radiation and the temperature is usually given in degrees Kelvin (absolute temperature). Most of the light sources which we encounter in which light is emitted because of incandescence have energy distributions approximating that of a blackbody. They vary from a candle at around 1800 K through tungsten lights in the range from 2700 to 3500 K up to zenith blue sky on a very clear day which may exceed 10,000 K. Sunlight departs considerably from such a distribution because of atmospheric absorption but can often be considered as being around 5500 K. A word of warning is desirable at this point. Because this series of energy distributions is so convenient it has given rise to the designation of light sources by their "color temperature." This designation does *not* define their energy distributions. It means that they have the same visual color as the indicated member of the blackbody series. We shall see in the next chapter that it is meaningless with respect to the distribution involved.

Light rays pass freely through space without interfering with each other except under very special circumstances. An illuminated space can thus be thought of as completely filled with light traveling in all directions until it encounters matter. This light is invisible to an observer unless it actually enters his eyes. It then is converted to some other form of energy that produces the response we call "seeing." Light can also be detected and measured if it is allowed to fall on a surface at which it is converted into some form of energy that can itself be measured, usually electrical energy.

Light, of course, follows the law of conservation of energy so that when it encounters matter all of its energy can be accounted for. Basically, as noted, it is either reflected, transmitted, or absorbed; but the absorbed light may be converted into heat, may release electrons, or may, in some cases, be reemitted as light. Such reemission is called "fluorescence" and this phenomenon is of considerable importance in the general subject of color. Such reradiated light is usually of longer wavelength than the "exciting" wavelength (Stokes' law) and hence may change wavelengths too short to be seen (ultraviolet) into visible light, but cases are just as numerous in which both are within the visible range.

Any material or device which makes radiant energy available in another form is called a "receptor" and the nature of the eye as a light receptor is of immediate concern to us. In the cases of familiar devices such as photocells, photomultiplier tubes, and so on, it is usually possible to consider the receptor itself as having fixed properties which are the same over the whole area of its usable surface. The eye as a receptor is quite different from this and the differences play a considerable part in normal perception.

THE EYE AS A NONUNIFORM RECEPTOR

The human eye is a roughly spherical organ, set in a socket in which it is free to rotate in all directions. At the front a compound lens (cornea and lens proper) is set into an opening through which image-forming light enters the eye. In young observers this lens is of variable focal length and changes, essentially without conscious effort, to focus objects at varying distances so that an image is formed at the back of the eye. Immediately in front of the lens is a variable-area aperture called the "iris," again working spontaneously and changing the amount of light entering the eye by a maximum factor of about 16 to 1. It tends to be wide open at low light intensities and closed down at high intensities. It plays little part in color perception except perhaps as it modifies lens aberrations. The lens itself does not pass light of the shortest wavelengths and is largely responsible for the termination of response at this end of the spectrum. As age increases the lens "yellows," increasing the absorption in the blue region and so tending to increase the wavelength of the shortest ones that can be seen. The enclosure behind the lens is filled completely by a watery substance called the "aqueous humor" that has very little light absorption.

Covering the entire rear surface of the internal cavity of the eye and subtending a solid angle of more than 180° at the lens is the light-sensitive area called the "retina." This consists of a very complex layer of nerves, nerve endings, interconnections, and light sensitive receptors. We need to note here only a few facts about the arrangement of this layer, although the details are of great importance for any theories as to how the eye responds to light. One of the problems is that the "hook-up" is so complex that it can appear to support almost any hypothesis as to nerve interconnections, "feedback," and so on. We might note that the actual light-sensitive elements of this array are buried deep inside it, with all of the elements that carry the message to other parts of the brain lying between the receptors and the light. The light is thus somewhat further modified by this tissue but, again, this fact can be lumped with general sensitivity characteristics, since it is customary to assume that such effects remain constant. We do need to consider, however, the lateral distribution of the retinal elements because this has an important bearing on perception.

There are two kinds of light-sensitive elements in the retina, known as the "rods" and "cones," distinguishable histologically by the way in which they are connected to the brain. The total number of these elements is of the order of 100 million, of which some 6 or 7 million are cones. There are, however, only some 1 million fibers which carry the responses of all of these out of the eye through the optic nerve. There is thus much grouping of the sensitive elements into single channels, and both this grouping and the distribution of rods and cones change systematically over the retina.

When we look directly at a small area in the field of view the images of this area fall at corresponding points of both eyes at a region called the "fovea." This is the region of sharpest vision and the most important part of the retina from our standpoint; so it is convenient to consider the above distributions from this point outward. The central part of the fovea consists almost entirely of cones, nearly all of which are connected individually to optic nerve fibers. These cones are also packed more tightly together here and the structure above them is much thinner, so that there is actually a depression in the retina in this region. There is thus a sound physiological basis for the superiority of detail perception in this area and when we say that we "look directly at" something we mean that we make the image fall here.

This rod-free area extends outward to around 2 or 3° in the external field. As we proceed further rods are found mingled with the cones and the percentage of rods increases steadily with distance. At the same time both rods and cones show a tendency to connect in groups to single nerve fibers, the tendency being much stronger for rods, and these groups become larger the greater the distance from the fovea. Vision for detail thus decreases steadily as we go outward.

Partly overlapping the fovea and surrounding it out to around 10° is an irregular, diffuse ring of yellow pigment known as the "macular lutea." This pigment may be partially crystalline and birefringent and so may account for the "Haidinger brushes" mentioned earlier. Its importance in color perception comes from its absorption of blue light, thus changing the spectral energy distribution of the light reaching the receptors that lie under it.

Rods and cones differ basically in the minimum intensity of light to which they can respond. This difference is due to the presence in the rods, at low light intensities, of a photosensitive pigment known as "rhodopsin." This material is very easily bleached by light and is assumed to thus produce a response in the rods in much the same manner as a photographic sensitizer in photography. The visual response so produced is essentially without color, and the sensitivity of the eye as a function of wavelength at these intensity levels corresponds to the wavelength-absorption curve of rhodopsin. It is distinctly different from the wavelength response curve of the whole eye at higher intensity levels, notably in its lack of response beyond about 650 nm. Vision at these low levels is called "scotopic." Because of the absence of rods in the fovea there is no response from this region for light of these intensities; to obtain a response it is necessary to look off to one side of the stimulus. Because grouping of receptors is the rule rather than the exception for rods, and the size of the groups increases rapidly outward, such vision is very indistinct and serves largely purposes of orientation and the detection of motion. We shall consider it further only incidentally. The mechanism appears to play little, if any, part in normal color perception at daylight intensities.

The fibers from the various receptors and groups of receptors cross the inner (vitreous humor) side of the retina and pass out through the retina together in the optic nerve bundle. This area is called the "optic disk" and is completely blind. Its area is comparable to that of the fovea and lies about 16° toward the nose in both eyes (outward in the external field) so that corresponding parts of the visual field cannot fall on both simultaneously. Even with one eye closed, however, an observer cannot become aware of this area except by making something disappear by arranging for its image to fall wholly on this area. The reason for this is not clear, except for the obvious fact that this area is not represented at higher centers in the brain. It is, however, the subject of a delightful hypothesis by Walls (1954), called the "filling-in hypothesis," to the effect that any *uniform* area surrounded by a *single-edge* contour is seen *because of* the change at the contour and not by the stimulus inside it. It is, I think, completely unprovable! Incidentally, those interested in how the human eye "got that way" could profit by reading at least the first part of his book *The Vertebrate Eye* (Walls 1942).

As intensity levels are increased above those for purely scotopic vision, color vision starts to be effective and there is a light-intensity region in which both mechanisms respond. Vision in this region is called "mesopic" and the range over which it occurs is relatively brief. Above this level and on up to the higher tolerable levels vision is said to be "photopic" and it is this region with which we shall be concerned. At these levels color perception is fully developed and the scotopic mechanism is at least assumed to be in abeyance. Under these conditions, however, the eye is not uniformly sensitive to color, the sensitivity varying outward from the fovea roughly in accordance with the distribution of cones. In good light color response can be considered normal out to at least 40° from the fovea. At the extreme center of the fovea, however, over a very narrow angle, the eye appears to be insensitive to short wavelengths (so-called "central tritanopia"). This may again aid detail vision by offsetting the rather severe chromatic aberrations of the eye lens.

The retina as a light receptor, therefore, is far from uniform over its area. But the importance of this for color perception is not so much the details of the nonuniformity as the fact that even a rather keenly introspective observer is not normally aware of them, unless they are called to his attention or he is puzzled by them under infrequent conditions. Perhaps the main reason is the phenomenon called "adaptation."

The sensitivity of the eye to light in terms of its visible response is anything but constant. In fact, probably less error is made by thinking of it as constantly changing than as static under any normal circumstances. The eye tends to act like what is known as a "null" instrument: it tends to respond more to differences in the field of view than to absolute values. It appears to do this automatically by adjusting its sensitivities to something approaching the average of the stimuli to which it is exposed, and it does this not only for the eye as a whole but also locally when the eye is held relatively stationary. Sensitivity is also affected laterally by stimuli lying near the one being considered. All of these are time-dependent, the time scale being determined largely by the magnitude of the change from the previous stimulation. Adaptation is also essentially independent in the two eyes so that they may be at quite different sensitivity levels simultaneously and, in fact, they often are.

The time factor is an important element in perception even under quite ordinary conditions. In general, adaptation is more rapid to an increased intensity than to a decreased one, but the time can be quite short (of the order of 0.2 sec.) for moderate changes in either direction. We are ordinarily not aware of these changes but we can become aware of rapid adaptation by a simple experiment. Simply close the eyes for a moment in a well lighted room and note the sudden decrease in brightness from the initial value that occurs just after the eyes are opened again. A similar change of much greater magnitude is commonly experienced in the sudden blackening of the windows at twilight when the lights are turned on in a room. For gross level changes we are all familiar with the painful glare of a bright light after a long period in darkness, adjustment sometimes taking as long as a minute, and of the utter blindness on entering a dark room from daylight, full adjustment then often taking as long as ½ hour. The total possible range of adaptation to intensity is not known but it certainly corresponds to a change of more than a 10,000 to 1 in average sensitivity.

It would seem that these effects, added to the nonuniformity of the eye laterally, would confuse the viewer as to what he is seeing. They obviously do not; and the remainder of this chapter is devoted to a consideration of some of the reasons. They lie in the way an observer uses his eyes to perceive the world around him. I shall call it "seeing."

THE SEEING PROCESS

The fact that the eye has a lens that forms an image on the retina leads at once to an analogy to a photographic camera, and it may be true that knowledge about the eye led to the addition of a lens to the "camera obscura" to improve its image. The analogy, however, is very misleading in that it suggests that the mind sees the image so formed.

A somewhat better, although still misleading, analogy is to a television camera. In this the scene is scanned constantly, point for point, at high frequency. Thus its image is transient and constantly being renewed. In a way this resembles the rapid "saccadic" movements of the eye that make it constantly traverse boundaries and, to some extent, the frequent blinking of the eyelids. Modifications of the image in the circuitry in television, particularly the so-called "crispening" circuits, may also have close parallels in vision, but little is gained by pursuing the analogy.

The purpose of vision is to contribute information about the external world to the brain. Having learned from experience, the observer uses vision in a way that will maximize the wanted details and minimize the unwanted ones. The skill of the average person at doing this is so great that it makes any attempt to describe how it is done sound like an exaggeration. We say, ordinarily, that a person's "attention" is concentrated on some particular phase of a scene and yet think of him as seeing everything else at the same time. As a matter of fact he does not. The situation is described nicely by what is said to be an old Arabian proverb, "The eye is blind to what the mind does not see." One of the defects of photography is that, except in the hands of a true expert, it does not have this power to select the important part of a scene. A person can walk into a room full of people and see essentially nothing but the face of a person 20 ft away. We say he has "come to see" that person and he does so literally. On the other hand, he may have come "just to look around" and may do just that. Vision is strongly selective and guided almost entirely by what the observer *wants* to see.

If the eye is deliberately held stationary in a new situation (a very difficult process), very little is really seen except within a few degrees of the direction of the eyes. Thus when we are talking to a person at a distance of 2 or 3 ft we cannot tell, except in a general way, what he is wearing. In order to do so we should have to "look him all over," something that we would not normally do unless dress were the subject of conversation.

The analogies with photography and television break down because it is *not* the image formed by the eye lens that we see. The image of a room, or any other situation, is in the mind. Any details in that image, except the very broadest, have been built up by successive glances directly *at* these details; both the details and the broad image are retained by the mind for as long as they are wanted and then as quickly erased. The optical image is constantly changing and moving as the eye jerks rapidly from one point to another; the mental image is stationary for stationary objects regardless of the eye motion, or, for that matter, the motion of the head or of the person.

Once we realize that essentially all the eye itself does is supply this detail to the mental image, then we can begin to understand how its optical and physiological properties can be utilized to enhance this information. As long as vision is thought of as formation of an optical image, the eye is a very defective instrument. Thought of as a mechanism for collecting and verifying bits of information to fill out an existing image, it becomes fascinating in its possibilities. Probably not enough is yet known for us to realize all that occurs but we can review some phases of it here and more later; they will help us understand what is involved in color perception itself.

We have noted that when we look directly at a point in a scene the image of this point falls on a part of the retina with little sensitivity to short-wavelength light. Outwardly from this point for a few degrees the image lies on an area of maximum resolution for detail and color. It then encounters a diffuse yellow ring where its energy distribution is modified by the absorption of part of the short-wavelength light. Beyond this area detail discrimination deteriorates fairly rapidly but color perception persists at normal light-intensity levels. Furthermore, when we first looked at this point the eye underwent, more or less instantly, three changes in its sensitivity to light-intensity. Considered broadly, it adapted generally to something like the average of the scene as a whole, it adapted locally to the area around the point, and its response to the point was modified by the characteristics of the adjacent areas. In somewhat the same manner but independently, as we shall see later, it also adapted to something like the average color of the scene as a whole, and this was again modified both locally and laterally. All of these adaptation processes have the effect of *maximizing* the detail and color differences in the scene.

As we look about a scene, rather than at a fixed point, the image in the eye moves about across the region of sharpest seeing as well as all the others. This voluntary, although not usually conscious, movement corresponds to shifting focus of attention on details. During each pause there is also a fairly rapid "saccadic" tremor of the eyes. Both movements have the effect that contours in the image are constantly crossing the receptor elements of the retina. Not too much is known of the actual effect of this but it is certain that it plays a large part in contour perception and it appears to be a necessary condition for vision. We shall consider it again in Chapter 12 when nonuniform stimuli are discussed briefly.

Perhaps the most important fact, in terms of perception, about the nonuniformities of retinal sensitivity, is that the observer is unaware of them and yet uses them to his advantage in getting the facts of the scene before him. Instances will become apparent in later sections but we might note a few examples here to indicate the sort of phenomena involved.

The yellow ring around the fovea, the macula lutea, is not normally visible, principally because of color adaptation of the underlying receptors. Another contributing fact is that it has diffuse rather than sharp boundaries. The underlying receptors normally have sufficiently increased short-wavelength sensitivity because of adaptation to offset the absorption. While the function of this pigmentation is, presumably, to decrease the quite large chromatic aberration of the eye, because of the enhanced sensitivity it is not obvious how this helps. The change in the spectral distribution of the light, however, means that some differences between colors that would not otherwise be observable can be seen, and it plays a very considerable part in increasing the number of colors that are distinguishable.

The short-wavelength insensitivity of the extreme center of the fovea may play more part in the perception of fine detail than is usually realized. I was surprised one time when giving a lecture to note that the black lines of a diagram being projected on the screen near me were a pure, rather light blue, apparently due to an aberration in the projector lens. Checking afterward I found that from as near as 8 ft from the screen these lines appeared completely black, the blueness apparently completely removed by the central properties of the eye.

The outlying areas of the retina play a very interesting but not very clearly understood role in color perception. As noted earlier, it is apparent from their nerve structure that they serve primarily to detect motion and to give broad orientation and perhaps aid in the stability of the mental image. In connection with the mind, however, they play a considerable part in what I shall call, for want of a better term, the "field of attention" in vision. In effect the eye-mind combination works at varying magnification, the field of attention varying from just a few degrees out to almost the limit of vision. The subject is elaborated in the little book *Scenery and the Sense of Sight* by Cornish (1935), which well repays reading. In photography this change in magnification is the equivalent of a change in focal length of the lens, although in the eye it is, of course, a purely mental operation. Changing the field of attention from that of a broad landscape to some detail in this landscape is the exact equivalent of the use of a "zoom" lens in motion picture photography. It is a frequent source of disappointment to amateurs to see the result of photographing a scene with a distant point of interest, perhaps a church with its steeple, that they wanted to record—it may be hardly visible in the picture! I once calculated the range of focal lengths that would be necessary to imitate this variable "magnification" of the eye. It came out several *hundred* to one.

Thus when the attention is directed toward a wide angle of view, the outer parts of the retina are in fact used in the direct act of seeing, but details would be supplied by a quick glance directly at them. It is probably the poorer vision of these outer regions that has given rise to the notion that memory color of objects plays a large part in color perception. It is not likely that it does. But in a broad scene an observer can easily be aware of, for example, a red house off to one side and *if it does not attract his attention at the time,* he will think of it, not as of its true color, but as of the color he associates through experience with red houses, that is, his memory color of red houses in general. If he had looked at it directly this would not have been the case.

SUMMARY

It is apparent that the physical light stimulus for color as it enters an observer's eye can always be specified in quite simple physical terms. We shall see in the next few chapters that certain (relatively fixed) characteristics of the response of an "average observer" can be used to simplify and arrange the infinity of such possible light stimuli.

The way in which the eye sensitivities and capabilities are actually used by an observer, however, is a complex mixture of intentions, desires, and interests that tend to preclude anything but broad generalizations. In spite of this we can go far toward understanding most of the color perceptions of a normal observer by considering what he sees when the light stimulus is presented in more and more complex situations, the physical characteristics of which are also known. In this way, starting with the simplest possible stimuli, we eventually arrive at the perception of color stimuli in everyday surroundings. This is the plan of the rest of the book except for the final chapters, which will apply what we have learned to a few special problems.

We should bear in mind in this process that this approach is the opposite of the lifelong experience of every observer. Born into a visually complex world, each person gradually learns to sort out what he sees by means of *all* his senses, with vision gradually taking the place of the others for objects at a distance. Fortunately for our subject, color is a response of vision only. For this reason we can hope that the various perceptual processes we uncover as we gradually synthesize the complexity of a natural scene will correspond closely to those involved when the observer, in fact, starts with such a scene and analyses it into its components.

Three *METAMERISM: PSYCHOPHYSICAL VARIABLES*

In spite of the evident variability of the eye as a receptor, there are certain relations between the stimulus and the visual response that are of sufficient general validity to be considered as "laws" that the eye follows. Such relationships are called "psychophysical" (although the term is often used broadly to cover the whole subject of stimulus-response studies). From these "laws" it is possible to deduce certain combination variables that, because of their general applicability, can be used to categorize the stimuli themselves. Because of the nature of the relationships it is found that this greatly simplifies description of the stimuli. Specifically it is found that all possible stimuli can be described in terms of three properly chosen psychophysical variables, of which many sets are possible. The empirical facts from which such sets may be derived are the subject of the present chapter.

METAMERISM

The analytical ability of the eye does not extend to seeing spectral energy distributions directly. The response to any distribution seen by itself is (usually) a unitary one that gives no inkling of the wavelength mixture causing it. It is not too surprising, therefore, to find that the color that an observer sees from a particular stimulus can be matched exactly by a great many other stimuli with entirely different spectral energy distributions. Such matching stimuli are called "metamers"; they are said to be "metameric" to each other, and the subject of matching stimuli is called "metamerism." The term is also used comparatively; two metamers whose spectral energy distributions are very different are said to be "strongly metameric," those with small differences, "weakly," and so on. An absence of spectral differences between matching colors may need to be pointed out by such a term as "spectrally identical," or "isomeric."

The eye is unique among the sense mechanisms in having a fixed initial response to its stimuli that is sufficiently predictable for a given individual and sufficiently alike in different individuals so that a whole science can be constructed around it. (A possible exception is the receptor for the sense of smell but it remains to be demonstrated.) The entire basis for the subject of colorimetry is the existence and nature of metamers and metamerism and the psychophysical variables that can be deduced from them. Colorimetry will be sketched in the next chapter insofar as we need it for our purposes, but the very ease with which colorimetric relationships can be reduced to mathematics tends to obscure the truly remarkable visual phenomena that are involved. Accordingly we want to consider first the characteristics of metameric pairs, then the ways in which metameric pairs can be formed, and then the relatively simple variables that emerge from these facts.

Suppose that an observer has in front of him two stimuli that are in close contact with each other, and between which he can see no color differences at all, but that are known to have distinctly different spectral energy distributions. We want, essentially, to list what is known about such metameric pairs.

We note first that not all observers would agree that they are a perfect match for color. Metamers, strictly speaking, are valid only for individuals. In fact an observer's color vision may be specified by the nature of the distributions that he does see as the same. Fortunately, however, these individual differences, even when large, do not modify the rules which such pairs follow, for stimuli which match *for him.*

By far the most important characteristic of such a pair is the essential invariance of the *fact that they match each other.* Within limits, the match as such persists even though the conditions under which they are seen may cause their color to vary throughout a considerable part of the range of observable colors. It is this invariance that both permits us to characterize stimuli by reference to a standard set of stimuli and at the same time warns us that such a characterization describes not the color seen, but only what it will match. Properly speaking then, we should refer to such pairs as "stimulus metamers" and not "metameric colors"; and while we will find later that such a distinction is in fact necessary for complete generalization, we shall not make it at the moment.

The invariance of metamers, so far as we will consider them, is coincident with the range of photopic color vision. It does not persist down into the mesopic range of twilight vision and it tends to break down at very high intensities. Thus it holds over the entire range of ordinary "daylight" vision and is independent of both color and intensity adaptation provided the spectral energy distributions of the stimuli themselves are not changed.

Perhaps the most remarkable fact about metamers is that they may, in many ways, be considered as identities. In any situation that does not modify its energy distribution, any metamer may be substituted for any other without visible change. Furthermore, metamers act as identities both in mixtures and on multiplication or division. Thus if we have a pair of matched stimuli, any color may be added equally to both without destroying the match, and an equal amount of any of *its* metamers may be substituted on either side. It follows that both of the original stimuli may be divided or multiplied by a constant without changing the *fact that they match.* In general, any such change will of course modify the perceived color of *both.*

These facts are basic to an understanding of color perception, and an analogy may be helpful for some, even though analogies are often dangerous. A metameric pair may be thought of as equal weights of quite different materials, balanced on the two sides of a pair of scales. Within the lower sensitivity limit and the upper load limit of the scales, it does not matter what either the actual weights or the materials are. We are concerned only with the fact that they balance. It is apparent that we could multiply or divide both sides by a constant, add equal weights of anything to both sides, and so on, without affecting this. These operations are analogous to the facts about metamers mentioned above.

We can carry the analogy further as an introduction to what you can *not* do with metamers without destroying the match. In the case of the balanced weights, it is obviously possible to maintain the balance by removing equal weights from each side, but only *if you know how to do it.* Suppose we have on each side of the scales a mixture of three sizes of small spheres (shot, for example), but all six sizes are different. We do not know the relationship of the sizes; all we know is the total number of each size, that is, the size distributions. Thus we do not know how to *subtract* from each side without destroying the balance. We cannot take an equal total number, nor an equal number of each size, nor an equal volume, from each side. We can, however, take away an equal percentage of each size, since this is the equivalent of multiplying each side by a constant fraction, but the *same* fraction must be applied to all *sizes.* We obviously cannot, for example, take away 10% of the larger, 30% of the smaller, and 20% of the intermediate size from each side. It is the equivalent of this latter case, however, that occurs with great frequency in color.

When colored light is reflected from a colored surface or transmitted by a colored object, the light is in effect multiplied by a fraction that is different for each wavelength. Thus a varying percentage of the light is removed, depending only on the wavelength and not on the amount present. This is the process of "selective" reflection or transmission characteristic of all colored objects. When objects reflect or transmit *equal* percentages of *all* wavelengths they are said to be "nonselective." It is apparent from the analogy, if not otherwise, that if a pair of metameric stimuli is subjected to selective modification the match will be destroyed, but this will not be the case if the modification is nonselective. Unfortunately such a selective modification has come to be referred to as "subtractive mixture" of the two colors. This is doubly unfortunate since it is neither a mixture nor subtractive in the mathematical sense; "divisive" would be a better word if its connotations were purer. The term, however, has a reasonable origin, and it is not out of place to consider it here. If two colored pigments are mixed, each can be considered, roughly, as continuing to selectively absorb its own share of any incident wavelengths. "Mixture," therefore, refers to the pigments and "subtractive" means they take away energy. Physically, the same thing occurs if light passes successively through two colored media or if a beam of colored light passes through or is reflected from a single medium, and the term has thus been extended to these cases. Since we will not be concerned with the physical origins of light distributions, I will have little need for the term itself, but the concept is encountered frequently.

A pair of metameric light stimuli, then, will cease to be metameric if both pass through or are reflected from a selectively absorbing medium before they reach the eye but will not be thrown out of balance by a nonselective medium. We can now, if we like, extend this to pigments. Two metameric pigments will cease to match if each is mixed with a third colored pigment *or,* what is essentially the same thing, if the light by which they are illuminated is changed to one with a different energy distribution.

Having thus considered what can and cannot be done to metameric stimuli, we can now return to more general considerations.

All possible spectral energy distributions that exactly match a given color when placed next to it constitute what I shall call a "metameric set" of stimuli, and such a set may be specified by any convenient member of that set. Such sets, however, differ widely in the number of possible distributions they may contain; the reasons will become more apparent presently when we consider the mixing of stimuli from different sets. In any given set, of course, the number of visibly different distributions is limited by the color discrimination of the eye; but the actual number may vary from very large for nearly or wholly colorless stimuli, down to a single value for some regions of the spectrum. The total number of discriminable *sets* of metameric stimuli has been estimated by Judd to be in excess of 10 million.

The existence of metameric stimuli with their invariance, interchangeability, and miscibility is thus one of the foundation stones for the science of colorimetry. Knowledge of the existence of metameric sets, however, is not sufficient; rules must be known for predicting the result of the mixing of stimuli from different sets or, more importantly, for giving a unique designation for a particular set. These rules, discovered and developed by many workers, culminating in Helmholtz, involve the various ways in which two or more stimuli taken from appropriately different sets may be mixed to produce a metamer for a given color. The rules were stated by Helmholtz as due to Grassmann and have come to be known as "Grassmann's Laws." We can note their chief characteristics here and give more details later.

If two colors from different sets are mixed (added to each other) in such a way that the total light remains constant, the appearance of the mixture changes continuously from that of the first at 100% to that of the second at 0%. In this series of mixtures, considered from the standpoint of color perception, the colors of the individual components may or may not remain recognizable. There are three limiting cases. First, they may both remain identifiable throughout as in the case of mixtures of red and blue to form the purples; in all of them both red and blue may be seen. Second they may, at some ratio, form a color that is unique in that neither color may be seen as such. For example, a mixture of orange-red and green can produce yellow. Third, and most important, at one point in the series the mixture may be colorless. Such pairs are known as "complementary" stimuli.

The second rule that we will note is that for any given stimulus that is not itself monochromatic there exists a number of pairs of sets of stimuli that can be mixed to exactly match it. This number may be quite small for colors that are nearly monochromatic and approaches infinity for those that are themselves colorless or nearly so. As a broad generalization, any stimulus can be exactly matched by a mixture in which one component can be chosen arbitrarily, the choice determining a whole series of other stimulus sets that with it will produce the match. Such pairs can also be said to be complementary *with respect to* the stimulus being matched.

Of particular interest to us at the moment is the special case in which the arbitrary choice is that of a stimulus seen as colorless. For this case any stimulus can be matched by some mixture of this with a specific monochromatic light or, for the purples, with a specific mixture of long- and short-wavelength monochromatic lights. It is apparent that such a pair affords a means of identifying the set to which a particular stimulus belongs since it is necessary to state only the amount and ratio of the two (or three) components. Attempts to use this as a method of designation, however, encounter a number of obstacles, and colorimetry has developed along different lines. It is of interest to consider these obstacles because of the insight they give into the nature of the problem and because of two other facts. The first is that this way of considering a stimulus is most closely aligned with the perceptual variables and the second is that this is the only *simple* way in which *all* stimuli can be exactly matched.

The main problem in the use of this method of designating a stimulus lies in the specification of the stimulus that is to be considered colorless. This is awkward to discuss at this point because "colorless" is obviously a perceptual term and not yet considered, and because it is not a unique property of any one set of stimuli: almost any set may be made to have this appearance under the proper conditions. It is perhaps sufficient to say at this point that the choice for this stimulus has to be made arbitrarily. We are fortunate that this choice has been the subject of international agreement, and we can hence consider this part of the problem solved. For the moment I shall call it the "achromatic point."

The second problem, already mentioned, lies in the fact that not all possible colors are represented by single monochromatic regions of the spectrum, the purples requiring a mixture of long and short monochromatic lights. As far as color perception is concerned, these mixtures, if the wavelengths are from the extremes of the spectrum, may be considered monochromatic, but their physical designation is awkward. All of them, however, are complementary (with respect to the chosen achromatic point) to a monochromatic wavelength which *is* in the spectrum, and it is customary to designate them by means of this wavelength.

Using these means, any stimulus (and hence its metameric set) may be identified by stating the wavelength, the percentage of this wavelength in mixture with the achromatic component, and the total amount found necessary to match it. The percentage of the monochromatic component is called the "colorimetric purity" and given the designation p_c. The wavelength, or complementary wavelength, is called the "dominant wavelength" with the designations λ_d or λ_c. Dominant wavelength and colorimetric purity thus uniquely define a metameric set but only in the sense that such a mixture would match any stimulus from that set *if* they were brought to the same total intensity.

In order to make these concepts quantitative, it is necessary to consider what is meant by an amount or an intensity of a color stimulus. This is the subject matter of photometry. However, it might be well to note that what we have done thus far is to find one method by which any color stimulus may not only be described but also actually produced. There are many other ways in which both may be accomplished and, more important, ways in which a given spectral energy distribution may be assigned directly to its proper metameric set. These are the subject matter of colorimetry and are taken up in the next chapter.

TERMINOLOGY

Up to this point we have not had to concern ourselves much with perceptual terms because we have been discussing mainly the empirical rules followed by mixtures of color to produce identities, and in identities all perceptual variables are matched. In order to go beyond these rules, however, and, in particular, to learn how to assign an arbitrary energy distribution to a specific metameric set, we need psychophysical "laws"; and these involve established relationships between physical and perceptual variables. For the moment we can define a perceptual color variable as any one of the *kinds* of perceived differences that can be seen when two colors are placed side by side, leaving a definition of "color" until later.

We can start consideration of these psychophysical relationships and the terminology involved by dividing all possible perceived colors into two groups, "chromatic" and "achromatic." The achromatic colors are those referred to earlier as "colorless." The chromatic colors are those in which the perceptual variable "hue" is present. In achromatic colors this variable is completely missing and is often referred to as having the value zero, although what is meant is that the color has no assignable dominant wavelength; that is, it has a *purity* (p_c) of zero. The variable, hue, is a basic sense perception and hence undefinable except by reference to common experience. It is the perception to which we refer in such words as "blue," "green," and "red," used without modifying adjectives. It is also, very often, what is meant when the word "color" is used as a combining form as we have above in "colorless." It can also be defined, loosely, as the principal perceptual variable that changes when the dominant wavelength *only* is changed by a small amount in a given stimulus. Note that this latter is an operational definition. In the last analysis this is probably the only kind of definition possible for perceptual variables, but such a definition has two obvious hazards that must be avoided. First, it implies that the physical change made to vary the perception has changed only the one variable. Second, it tends to identify the perceptual change with the physical one. The first has led to an unwarranted simplification of perceptual variables and will be discussed in detail in Section II. The second leads to what is known in psychology as the "stimulus error": confusion of the response with the stimulus, often resulting in the suggestion of a relationship that may not exist. A case in point is the present division of perceived colors into chromatic and achromatic. These are perceptual terms properly applied only to the response of the eye to a stimulus and *not* to the stimulus itself. The same stimulus that is seen as achromatic under one set of conditions may be made to produce all of the hues by changing its environment. The expression "achromatic stimulus" is thus misleading unless it is always understood that the proper environment is implied; there is no warning for the naive in the phrase itself. A good example of the type of error involved is the almost universal use in the literature of the term "white light." There is no such thing! What is meant, of course, is light that appears achromatic, but the term implies that there is a stimulus that always so appears, which is not true, and it makes "white" synonymous with "achromatic." It also leads to unnecessary confusion in the important subject of object color perception, where the word "white" requires a restricted meaning. On the other hand there are many instances where the use of such a term as "achromatic stimulus" is fully justified because of the simplification it introduces in discussing a subject. Our use of "achromatic point" above is an example. What is meant is *the* achromatic point of the system under discussion and avoids constant repetition of this whole phrase. I wish such instances could be printed in red ink; they have led in the past to many invalid concepts of color perception.

Incidentally, there are a number of words that are used in several senses so consistently in common speech and that so greatly increase the ease with which some subjects can be discussed, that there is little point in trying to restrict their definitions to a single meaning. A typical example is the word "color" and its opposite, "colorless"; another is the verb "to see." It seems better to let the context supply their meanings. I shall use "to see" occasionally to emphasize the restriction of my meaning to the ultimate mental perception.

We have also been using the words "amount" and "intensity" loosely in quantitative reference to light and need now to consider the psychophysical basis for actual units.

PHOTOMETRY

Suppose that we have two metameric stimuli presented as halves of a circular field small enough to be encompassed by the fovea and with no surrounding stimuli. This is typical of the viewing situations from which a large part of the data on color perceptions have come.

The fact that these two stimuli are metameric does *not* mean that each beam is sending the same amount of energy to the eye at any moment. One beam may, for example, contain a relatively large amount of energy at wavelengths that are invisible. I shall use the phrase "radiant intensity" as essentially synonymous with "rate of flow of energy" and, in general, use it only to indicate total energy. Thus when I say that the radiant intensity is increased I mean that the total energy has been increased without change in its wavelength distribution. The radiant intensities of the halves must thus be thought of as different although they match visually. Nevertheless I have said that metamers send exactly the same amounts of light to the eye, so we need a quantitative measure of "amount." The word "light" itself is restricted to visible radiation, but the eye varies in its sensitivity to different wavelengths within the visible spectrum. Briefly, the amount of light is the total radiant intensity evaluated with respect to the sensitivity of the eye. These are obviously connected by the efficiency with which the eye converts the various wavelengths to a visible form, and it is this relationship we need to establish. This relationship is known as the "luminous efficiency" function of the eye and is a psychophysical variable. It is as well established as the "color mixture" functions that lead to metamerism but is probably neither as stable with conditions nor as repeatable between individuals. To obtain this function on an absolute sensitivity basis we need a standard light source of known energy distribution and total energy; and to obtain the function on a relative basis we need a perceptual variable for which two such stimuli can be equated. The two together are the foundations for the subject of photometry, the measurement of light.

Photometry being a much older subject than colorimetry, the original light source is a standardized candle, long since replaced by a more repeatable and sophisticated source but still used as the name (officially "candela") of the basic unit. Thus we are free to think of the subject in terms of the light from an actual candle and can derive from it the few photometric terms we need. We shall be concerned almost exclusively with areas in the field of view that send equal amounts of light to the eye from each point, that is, with uniform areas. It does not matter perceptually whether the light actually originates from this area or whether it comes from some other source, nor whether it has been reflected or transmitted in order to reach the eye. The situation with which we are concerned can be represented by what is seen when the light from the candle falls on a diffusely reflecting surface. If we take this surface as reflecting 100% of the light without modification of its energy distribution and assume that all points of it are at a distance of 1 ft from the candle, we then have a quantitative representation of the unit called one "footcandle." This unit is properly used only in reference to the light falling *on* the surface. We are concerned primarily with the light which we see reflected *from* it. Because the appearance of this area is independent of its distance from the eye, and it does not matter what caused the amount of light to fall on it, a different unit is applied to this aspect. This unit is called the millilambert (mL) and I shall simply change to it. In this sense, one footcandle is equivalent to 1.076 mL. If we now disregard the way in which this particular area of light was produced, we can generalize this unit to describe all the uniform areas of importance to us, regardless of how they were produced, and state the amount of light in terms of millilamberts. Candlelight is, however, quite yellow and of low intensity, and the problem is to learn how to go from this unit to all colors.

Returning to our metameric pair of stimuli, we will assume that both have the energy distribution of candlelight and are at 100 mL. This places them well within the photopic range, although toward the lower end. If we now increase the amount of one of them by a few percent, the match will be destroyed, and we can consider the changes that have been introduced. First, and most important from our standpoint, is the change in appearance of the increased stimulus to a condition that can be described by the common speech term "brighter." This introduces the perceptual variable "brightness" and, in fact, serves as an operational definition of the term. Thus we can say that brightness is one of the perceived color variables and that this is the variable chiefly affected by a change in the amount of light. Note that two phases of the preception are involved, the absolute brightness level and the relative one implied by "brighter." There is also the brightness difference between the two. This brightness difference has been found to be approximately independent of absolute brightness over the photopic range and leads to the psychophysical relation known as the Weber-Fechner law: the brightness difference between two stimuli depends on their intensity ratio, if the energy distributions are the same. We need not consider it here other than to note its existence.

Second, we note that while the change has been specified in physical terms, and the brightness difference is amenable to a psychophysical law, we can say nothing about the absolute brightness except by comparison with another stimulus. Ultimately we have to express brightness by saying that a given brightness is that which is produced by so many units of the basic standard, and we want to be able to do this not only for stimuli that match in all respects but also for stimuli in which there is a large difference in dominant wavelength and purity. This is the subject of photometry, and we can describe it, for our purposes, as the science that deals with the brightness aspect of colors as distinct from their other perceptual attributes.

We might note here, as a third item about the change, that I have assumed brightness to be the *only* perceptual variable involved in the change of appearance. I shall present evidence in Section II that this is not true, but for the moment this need not concern us because we are considering only brightness matches.

We want to discuss brightness matches between stimuli of different colors and, particularly, to find a psychophysical relationship that will permit us to evaluate two different absolute energy distributions with respect to brightness.

As long as there is little chromatic difference between stimuli and one of them is continuously variable with respect to the other, there is little difficulty in setting a brightness match. The second can then be said to represent the same number of millilamberts as the first, and this process can be repeated until any desired chromatic difference is reached. This is in fact the basic operation of photometry. Any stimulus in the series can of course be used as a secondary standard for direct comparison with other stimuli of the same color. The basic standard can thus be extended usefully over a very considerable range. This range includes both the color and the intensity of the new standard (1948), a blackbody radiator at the temperature of solidification of molten platinum (2046 K), and the candle at around 1800 K.

In color, as distinct from simple photometry, we are concerned basically with the stimuli most remote from that of the photometric standard, namely, monochromatic light. We need to know both the relative efficiencies of the eye at the different wavelengths for the production of brightness and the way in which lights combine in this respect. Having established the equivalence to the standard of one wavelength, we can calculate it for the others and their mixtures and so evaluate the equivalent for an entire energy distribution.

We can dispose of the question of combination at once. When lights are expressed in the same units (mL) they simply add. This fact is sometimes known as Abney's law, and its universal applicability has been questioned. It is, however, accepted as a basic rule of colorimetry.

We can concentrate then on the first question, that of the relative efficiencies of the eye for energy at different wavelengths. When two monochromatic stimuli of widely different wavelengths are placed in the halves of a photometric field and an attempt is made to match them for brightness, the appearance is confusing; and the point at which a match for brightness only is reached is very difficult to determine. The results obtained are thus subject to large errors, and the same situation exists for direct comparison with a standard. For this reason many other methods of comparison have been tried, among them visual acuity and variation in pupil diameter, but only two were found that gave sufficiently consistent results: these are the so-called "cascade" and "flicker" methods.

In the cascade method monochromatic lights are presented in a comparison field, and means are provided for varying both the intensity and the wavelength of each. The wavelength of one light is adjusted to give a small hue difference between it and the other, and brightness is then matched. Starting with the second wavelength the process is repeated until the end of the spectrum is reached. The process is then reversed to reach the other end, and the whole spectrum is thus repeatedly scanned until sufficient precision has been obtained. The process is not entirely independent of the exact procedure and is poorly repeatable from day to day as well as being exceedingly tedious. It is also obviously suspect on the grounds of both systematic and cumulative effects. Nevertheless it is the method most workers have believed to give the truest result, and these results have been used as a sort of norm to check other methods. All observations, however, were consistent to the extent that they shared a maximum of sensitivity near the center of the visual spectrum and a curve shape similar to that of a slightly skewed Gaussian probability distribution, with no obviously repeatable departures from a smooth curve either side of the maximum.

The problem of finding a more repeatable method that was less demanding and so could be carried out by many observers whose results could be averaged became rather acute early in this century. Because of practical pressures, largely from the field of illuminating engineering, a great deal of effort was then devoted to finding such a method. This was eventually found and became the basis of a widespread method of equating heterochromatic brightnesses, known as "flicker photometry."

The flicker that is produced by alternating two different color stimuli can be eliminated at a lower frequency for the chromatic aspects of the colors than for the brightness differences. This fact suggested the possibility of a method for equating brightnesses that would not involve the necessity of judging brightness equality as such.

From the very start the method was suspect because of the well known fact that stimuli of brief duration could produce appearances very different from those seen in the steady state. Nevertheless the methods evolved did produce results that were more easily obtained and more repeatable than any other method known. We do not need to consider the details of the method, but it is important that we understand what is involved perceptually.

If, in a single viewing field, two different color stimuli are alternated, what is seen depends both on the frequency of alternation and on the differences between them. As the frequency increases, four phases of appearance are seen. At first there is a very unpleasant alternation of the two, each being seen separately. This is followed by a fairly critical range of speeds at which the chromatic aspects of the two tend to fuse into a single color which still flickers. If, at such a speed, the intensity of one of the stimuli is varied, it is found that this flicker may be reduced to a minimum or to zero, depending on the flicker rate. At this point the two stimuli have equal brightnesses, essentially by definition. If the frequency of alternation is too high, the stimuli fuse no matter what the brightness difference, and change of one of the intensities then simply changes the chromatic appearance of a visually steady field. Settings were found to be most critical and repeatable when some chromatic flicker was still present and the total flicker was set at a minimum amplitude by intensity changes.

This method had the great advantage that it could be carried out by relatively naive observers and so made available a larger quantity of data. The problem then shifted from that of getting an average curve at all to that of deciding whether the curves so obtained were the same as those that would have been obtained by less suspect methods. This was a somewhat simpler problem since comparisons between *methods* could be made by single highly skilled observers, and it could be assumed that the visual *mechanism* involved was essentially the same for all observers.

At first, such studies showed a wide diversity of results. Gradually, however, conditions were found that if repeated exactly gave essentially the same results as those obtained by the cascade method. Furthermore, given these results, it became possible for experts in the field to recognize the perceptual variable to which it corresponded and so learn how to make direct heterochromatic matches for this variable between monochromatic stimuli. Thus considerable confidence in the method developed, even though it was not obvious exactly what was being measured.

This work uncovered large variations among individuals, due not only to method and individual observer, but also to just as great an extent to the conditions under which the stimuli were viewed. It became evident that it was more important that *a* set of data be agreed upon than that it have a specific correlate. Such a set of data was produced and accepted as an international standard in 1924 (see the following chapter), and it was agreed to postulate a Standard Observer having these characteristics. It was also postulated that for him (or it) mixtures of energies of different wavelengths were strictly additive with respect to this characteristic.

The data thus standardized form, when plotted, what is known as the "luminosity curve" (for an equal energy stimulus) or, more properly, the "relative efficiency" curve for the Standard Observer. By means of this set of data any light stimulus may be evaluated, wavelength by wavelength, the sum (or integral if continuous) giving a single valued "luminance" for the stimulus. This can be expressed in absolute terms, relative to the standard source. More frequently such calculations are used in practice to evaluate the effect of selective reflection or transmission by colored objects. It is then expressed in percentage relative to a nonselective, nonabsorbing medium and called "luminous reflectance" or "luminous transmittance." It must be kept in mind that such calculations are based on psychophysical relations assigned to the Standard Observer. The justification for their use in practice lies in the fact that they "work." But we shall be much concerned with exactly what perceptual variable is involved in the metamers they predict.

With this in mind we can now consider in some detail the direct comparison for brightness of stimuli that appear very different chromatically. Basically, *why* is such a match so difficult that it can be made repeatably only by experts? It is customary to say that such a match is "confusing" or that brightness differences are "obscured" by the strong presence of other variables. Yet, if brightness is in fact a separate variable, why is it such a problem? The answer almost necessarily has to be either that there are one or more other variables present so much like brightness that it is difficult to distinguish between them, or that brightness in one hue is perceptually different from that in another hue, or that some new factor is operating in such a comparison. I think there is probably a little of each in the complete answer.

Parsons (1924), in his amazingly comprehensive review of the literature of color vision up to about 1920, quotes Helmholtz as follows: "I scarcely trust my judgement upon the equivalence of the heterochromatic brightnesses, at any rate upon greater and smaller in extreme cases." And, "As far as my own senses are concerned I have the impression that in heterochromatic luminosity equations it is not a question of the comparison of one magnitude, but the combination of two, brightness and color-glow (Farbenglut), for which I do not know how to form any simple sum, and which too I cannot further define in scientific terms.'' So far as I know this translation by Parsons is the only one we have, and the history of the passage itself is of some interest to us.

After the great success of the first edition of his *Treatise on Physiological Optics,* published in 1866, Helmholtz was asked by his publishers to prepare a revised edition which was published as the second edition in 1885. When W. Nagel was asked to prepare a revision for a third edition, bringing the material up to date, the decision was made to reproduce the first edition verbatim with added notes rather than to proceed from the second. Whatever the wisdom of the decision, it is only this third edition that we have in English and we are thus deprived of some of Helmholtz's mature thinking, including the passage above, taken from the second edition (page 440). It is interesting that this passage seems to have been ignored in the notes of the new editors. We shall find that it can act as an adequate introduction to the work described in nearly the whole second section of the present book, where the two perceptual variables involved are, in fact, separated and identified. It is unfortunate that I was not aware of the passage myself until undertaking the present writing.

At the moment, we want to consider the conditions under which the luminosity curve has been obtained. In the flicker technique the observer usually sees only a single stimulus area, although it is sometimes surrounded by an achromatic area of ''about the same brightness.'' In the cascade method he sees two adjacent areas that have a minimal chromatic difference. In both cases, therefore, the eye is chromatically adapted to the stimulus itself. In the heterochromatic case, however, the eye must either adapt to something in between the two stimuli or, more probably, change its adaptation slightly as the observer looks from one to the other. Furthermore, in the process of matching, the two variables involved do not respond in the same way to changes in the intensity of the stimuli. This fact, however, makes it possible to *learn* which of the two is the one measured by the other techniques and so *learn how* to make a heterochromatic luminance match. The experts could thus confirm the results of the other methods.

Confirmation of the luminosity curve came also from the discovery of a quite surprising fact that carried much weight at the time but has received little attention since. As an image, formed from an external stimulus, is moved across the retina away from the fovea there is a gradual weakening of the chromatic response. At extreme distances this response essentially disappears, leaving only a brightness response. The effect is usually ascribed to the decreasing percentage of cone receptors, to which chromatic response is attributed. It is found that in this region, where only brightness is visible, flicker matches and heterochromatic matches are in substantial agreement. We cannot consider this here from the standpoint of color vision theory but will only note that the luminosity curve holds true in the absence as well as in the presence of chromatic perception as such. We can include this with the previous statements and summarize both by saying that the brightness indicated by the luminosity curve is premised on the eye being *adapted to* the stimulus and appears to be independent of the chromatic response. We might also note that this response of the outer fovea is quantitatively different from the scotopic response at low light levels, mentioned earlier; so it may be helpful to consider this scotopic region a little further before taking up colorimetry itself.

SCOTOPIC VISION

As we saw it in the previous chapter, vision at very low light levels is due to an auxiliary mechanism that can be assumed not to operate at the higher (photopic) levels. The relative efficiency function for these low levels differs largely in the fact that the maximum efficiency is shifted some 43 nm toward the shorter wavelengths, without much change in the *shape* of the response function. This results in a marked increase in relative efficiency for short wavelengths and a very great decrease in efficiency for the long wavelengths. No hue is produced by any stimulus at these levels. The response mechanism is nonexistent in the fovea, and thus there is no response whatever at these levels; the observer is blind for scotopic stimuli falling on this area.

Although I shall restrict what follows to photopic vision, it is important to understand the perceptual effects involved simply because they are of daily occurrence. As the general level of light intensities in, say, a normal scene gradually decreases, as it does at "dusk," the sensitivity level of the eye increases by adaptation, reaching a maximum when night "falls," or soon thereafter. This maximum sensitivity is so high that it is a rare night indeed that is so dark that an observer cannot see the ground at his feet, although individuals vary greatly in their "night vision." It is often stated that under these conditions the eye is totally "color blind." While such a statement may be defensible as an explanatory technique, it is likely to be very misleading because it suggests that something has happened to the chromatic response mechanism. In fact when the eye is totally dark-adapted its chromatic sensitivity is at the maximum it ever reaches. It is the stimuli that are inadequate to produce a chromatic response. Any stimulus in the field of view whose intensity is sufficient, like a distant light source for example, will produce a chromatic response while everything else remains colorless. I emphasize this point because it is not so obvious that the inverse is true at levels where the photopic response is just beginning to *fail.* Under conditions in which hues can still be seen there may already be many visible stimuli that are inadequate to produce a hue response perhaps because of the nonuniformity of the illumination. In fact it is just this mixture of adequate and inadequate stimuli and of the appearance and disappearance of stimuli at the fovea that produces the "mysterious" effects of twilight and the like.

The mesopic range of intensities (also known as the Purkinje range) involves a gradual transition from photopic to scotopic vision. Over this range, metameric matches gradually break down for stimuli seen *outside* the fovea. All vision within the fovea can be considered photopic. Although the adaptation state of the eye plays a large part in what is seen in this mesopic range, especially if the transition is sudden, much confusion is avoided if it is thought of as a region of changing adequacy of the stimuli rather than as a changing condition of the eye.

We might note also that this discussion is an excellent example of the problem encountered in all discussions of perception. The terms "photopic," "mesopic," and "scotopic" refer specifically to the normal sensitivity states of the eye when adapted to high, low, and very low light intensities. These sensitivity states, however, are *caused* by the intensity levels, and it is not meaningful to refer to a *light level* as scotopic unless it can *also* be assumed that the eye is adapted to that level; again, of course, the "stimulus error."

In somewhat the same connection, it may prevent some later confusion if I mention here the perception of "black." Black is a contrast phenomenon, in the true sense of the phrase; it is produced by the contrast between two areas and not by a particular stimulus. It has nothing to do with scotopic vision even when the area seen as black is sending *no* energy to the eye. In fact, it is a perception associated almost entirely with photopic vision; a good black cannot be produced at scotopic levels, and the higher the photopic level, within limits, the better the black that can be produced. Even for moderate levels a stimulus seen as a good black could well appear blindingly bright to a fully adapted observer at scotopic levels. Roughly, black is the perception of the inadequacy of a stimulus area to produce a light response under the conditions of observation and has nothing else to do with the stimulus as such. It is no more startling scientifically than the production of an intense red by absorption of three-quarters of the spectrum!

Four
COLORIMETRY

We have seen in the previous chapter that photometry progressed from direct visual comparison of light sources, using absolute or derived standards, to the ability to calculate such comparisons. This was accomplished by the establishment of a psychophysical relation sufficiently characteristic of normal observers that it could be adopted internationally. It was agreed that any two light stimuli that calculated to the same value by means of this function would have the same "luminance" by definition and that, *for the postulated Standard Observer,* these two when observed directly, side by side, in a 2° split field would match for brightness. This function was to be used for all luminous intensities through the photopic range and for all adaptation states of the eye, but restricted to 4° or less foveal viewing.

This standard psychophysical function was adopted in 1924. The international group that adopted it was then reorganized into the present "Commission Internationale de l'Eclairage" (CIE), with the purpose of providing a similar basis for standardizing the measurement of color in addition to luminous intensity, that is, to add a specification of the chromatic aspects to that of luminance. It had long been known that this was possible; it required only sufficiently accurate data for agreement on further psychophysical functions of the Standard Observer. These are known as the "color matching" functions and were adopted in 1931.

Any set of three colored lights, no one of which can be matched by a mixture of the other two, and that taken together in proper proportion will mix to produce an achromatic stimulus, can also, in proper proportion, be mixed to produce all possible hues. The lights of such a set are called "primaries," in the sense of "coming first," the word having no other connotation when used in this connection. The purities of the colors producible by such a set are never greater than that of the purest primary but, of course, give all the other purities down to 0 for all dominant wavelengths. Any color within the gamut of a given set may thus be defined as matching a certain mixture of the three primaries. Since any color represents a whole set of metamers, this mixture of the primaries also designates the whole metameric set to which the color belongs.

Since the choice of the set of primaries is arbitrary we can assume the choice of a second set of primaries whose gamut includes that of the first (including the primaries). These second primaries will also mix to produce the same achromatic point. It is apparent that any color included in both gamuts may be defined by mixtures of either set, and, in particular, the primaries of the first set may be defined in terms of the second set. This operation is known as a "transformation of primaries" and is the basic operation by which the most useful set of primaries was finally chosen.

No such set of real primaries can match the purities of all the spectral colors because of the decreased purities of their mixtures. The outstanding contribution of Grassmann, Helmholtz, Maxwell, and others (c. 1860) was the demonstration that when the amounts of the primaries are expressed in proper units they may be treated as algebraic quantities. Thus, not only could any color within the gamut of a set of primaries be expressed as the sum of the required amounts of each, but colors *outside* of the gamut could be represented by the use of a *negative* quantity of one of them. Physically this meant that the primary was mixed with the actual color being measured, thus bringing it within the purity gamut of the primaries. In this way the entire spectrum could be "matched" by any set of primaries and expressed in terms of their amounts. The entire range of monochromatic stimuli, including the purples, could thus be described by a set of color mixture curves for any set of primaries; but over a greater or less part of the range these would show a negative quantity for one or another of the primaries. These negative quantities are unreal in any physical sense and are meaningful *only because* they have been demonstrated as legitimate. Accordingly, it was equally legitimate to arbitrarily assume a set of primaries that were equally unreal but had imaginary purities enough greater than unity to include all spectral colors and their mixtures within their gamut. These could be specified by algebraic transformation of primaries from any known set of color mixture curves of real primaries. This is in effect what was done, but with an added embellishment that greatly facilitated the *use* of the set in practice. This consisted in selecting the imaginary primaries in such a way that all the *lumi-*

nance of the specification for a given color was represented by a single primary, again legitimate because of the purely algebraic operations involved. It is not immediately apparent how this can be done, but it amounts simply to making the *color mixture* curve for that primary coincide with the previously adopted relative luminous efficiency curve for the Standard Observer. This gives a functional definition for that primary; its utility will become apparent presently. For our purposes the other two primaries were chosen simply so that the three together included the whole spectrum locus in their mixture gamut and so represented all possible colors by positive values of the primaries. Again, they are, in effect, defined by their mixture curves.

Since all the luminosity produced by mixtures of these primaries is assigned to one of them, amounts of the primaries cannot be given in physical units. They can be expressed in the *same* units, however, by stipulating that when mixed in an equal amount of each they match the achromatic point of the system, in this case the color produced by a stimulus having equal energies at all visible wavelengths (the "equal energy" source). This is the equivalent of specifying that the *areas* under each of the color mixture curves be the same.

The standard adopted by the CIE in 1931 and made part of the repertoire of the Standard Observer was three sets of data giving the ordinates of each of the above primaries as a function of wavelength, with the 1924 luminous efficiency curve as one of them. The functions represented were given the designations $\bar{x}$, $\bar{y}$, and $\bar{z}$, the $\bar{y}$ being the 1924 function. They are plotted in Figure 4-1. The sum of the standard ordinates for each complete curve is the same for each, meeting the requirement of equal areas. It may be helpful to think of each curve as a relative wavelength efficiency curve *of a primary* for a receptor belonging to the Standard Observer. It is used for calculation in the same manner as we noted earlier for the calculation of luminance.

Given the spectral energy distribution of a stimulus in either relative or absolute units, it is necessary only to multiply, wavelength by wavelength, by each of these curves, and summate or integrate separately for each function, to obtain numbers that represent the proper relative amounts of the three primaries to match the color. These three numbers were given the designations X, Y, and Z, the Y giving the luminance or relative luminance immediately, depending on the data and procedure used.

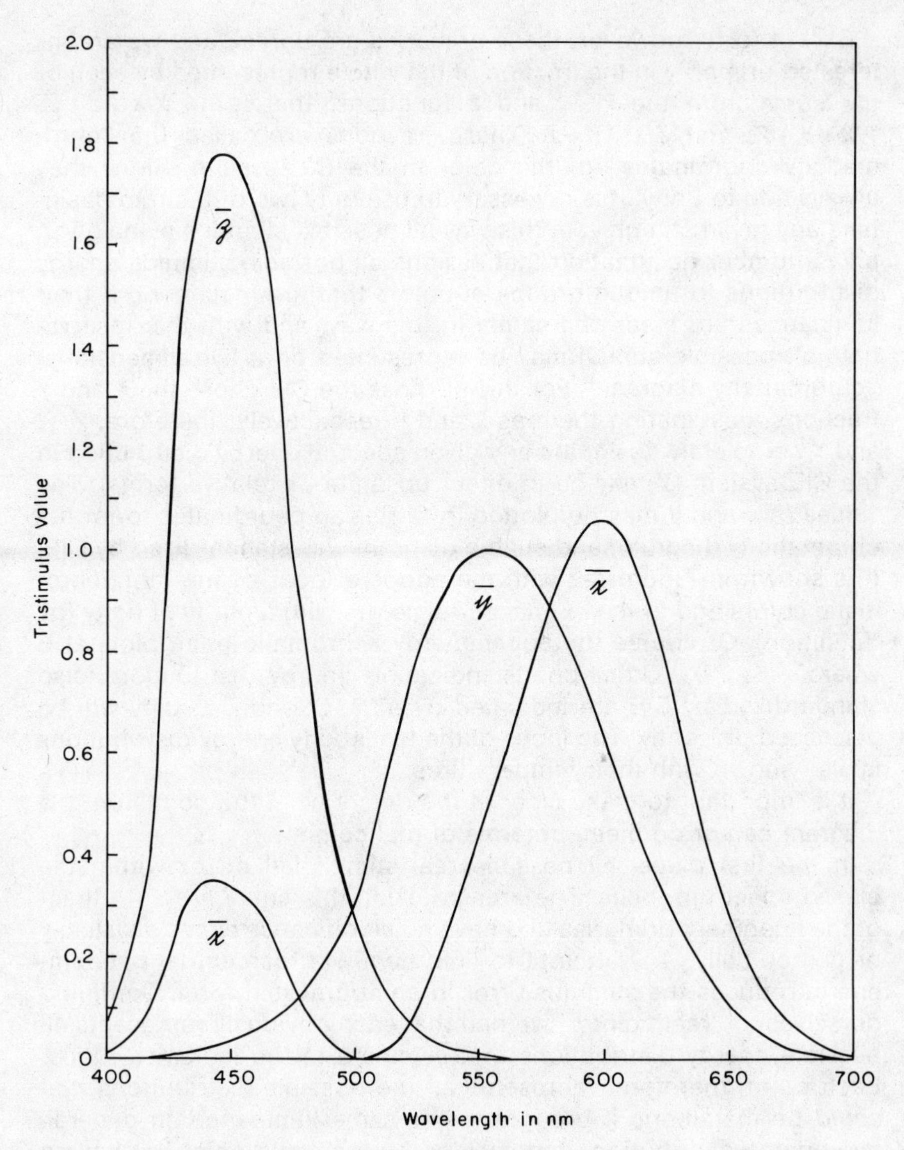

Fig. 4-1. CIE 1931 "color mixture curves."

Except for *Y*, however, these primaries are unreal, and we are interested primarily in the fraction of the whole represented by each of the sums from the $\bar{x}$, $\bar{y}$, and $\bar{z}$ functions, that is, in $X/X+Y+Z$, $Y/X+Y+Z$, and $Z/X+Y+Z$. These fractions are called the "chromaticity coordinates" of the color in the CIE system. Since they always add to unity, it is necessary to use only two of them to designate any given stimulus. In this way all possible stimuli may be given a two-number designation that assigns all possible stimulus energy distributions to unique groups of colors that are metameric *if* their luminances are made the same. In this way, and with this reservation, all possible stimuli may be represented on a two-dimensional "chromaticity diagram." For this purpose the CIE chose the *X* and *Y* fractions, designating them as *x* and *y* respectively; therefore, *x*, *y*, and *Y* completely designate any given spectral energy distribution in the CIE system (*Y* may be in either absolute or relative terms). The values of *x* and *y* may be plotted in Cartesian coordinates to form a chromaticity diagram, and such a diagram was standardized by CIE. It is shown in Figure 4-2 with the adopted locus of the monochromatic colors and of the extraspectral colors with a purity of unity (by definition). Of course the equal-energy achromatic point plots at E where $x = y = 1/3$. Other points indicating energy distributions also standardized by CIE are indicated by A, B, C, and D_{65}; they will be discussed presently. The locus of the blackbody energy distributions is also shown with their temperatures.

It is important to make clear at the outset what the points on this diagram can or do mean in terms of real colors.

In the first place, all possible real stimuli fall on or within the closed spectrum locus. The areas outside this curve are an artifact of the imaginary primaries and have no significance *either* physically or perceptually. Any attempt to link them with perceptual phenomena introduces the stimulus error in an aggravated form. Confining ourselves to "real points," we find that each obviously represents all possible energy distributions that calculate to the indicated *ratios*. Each point, therefore, represents all the possible distributions that *could* be metameric if brought to the same luminance. In general, two energy distributions represented by the same point but having different luminances would appear different, if seen side by side, not only in brightness but in hue and other characteristics as well. Thus the fact that two stimuli have the same chromaticity indicates not that they are metameric, a frequently encountered error, but that they may be potentially metameric. They are metameric only if the luminances are also the same.

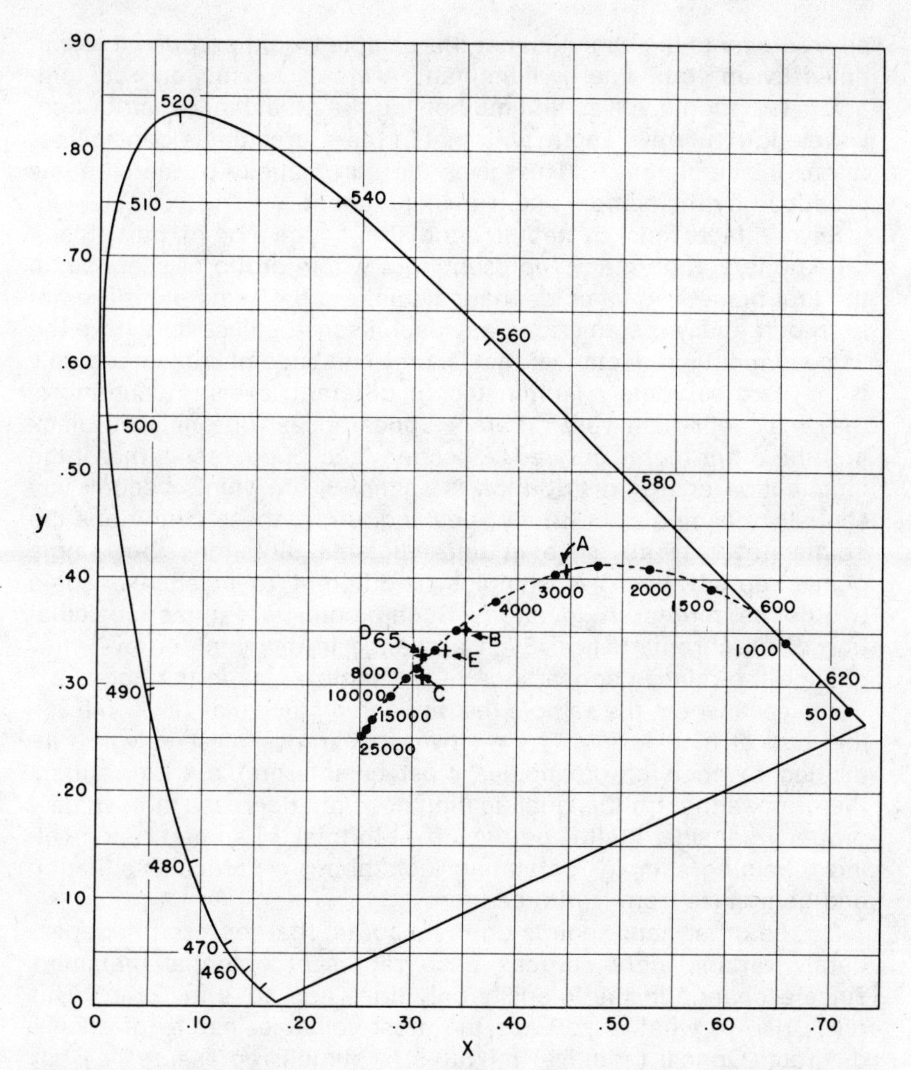

Fig. 4-2. CIE 1973 chromaticity diagram showing the blackbody locus and CIE stimuli A, B, C, D_{65}, and E.

The question as to whether or not two stimuli with the same chromaticity are potentially metameric, while fairly obvious for light sources, can become a source of confusion for other stimuli. For example, if the spectral reflectance (or transmittance) of a sample is known in terms of the percent reflectance at each wavelength relative to a 100% reflecting nonselective sample, a chromaticity point may be calculated directly from this information. Such a point actually represents the stimulus that this sample would produce if illuminated by an equal-energy illuminant. A more realistic, and customary, calculation involves first multiplying the reflectance distribution, wavelength by wavelength, with that of the source being considered, often CIE Illuminant C. This gives the chromaticity of the stimulus produced by the sample under this illuminant.

Several facts must be kept in mind about such a point, particularly if it is one among many representing a whole group of samples. In the first place, two samples that calculate to the same point are not even potentially metameric in any useful sense unless they have the same luminous reflectance; that is, they can become metameric only if they are *separately* illuminated to different levels. Furthermore, they may represent very different appearances; one may be black and the other highly colored. A second fact, of course, is that if the illumination energy distribution is changed, the points become essentially meaningless. Strictly speaking, under these conditions the points not only all move in unpredictable directions, but single points representing two samples of different reflectance become two points in different locations. Such a change requires recalculation of all chromaticities. Even a change in illumination level may change the relative appearance of the samples, as in the above extreme case where the sample that appeared black may take on a visible hue at high levels. The relative appearance of samples as indicated by their chromaticities must be interpreted with caution, therefore, although this qualification does not decrease their value if properly considered. It is perhaps best to think of a chart representing reflecting samples as stating facts about a particular situation and generalize from it with caution.

On the other hand, points on the diagram that represent independently variable *light sources* each represent potential metamers since a change in the intensity only does not move the point. This gives rise to what is perhaps the most valuable feature of such a diagram. Optical (additive) mixtures of stimuli represented by any two points always lie on the straight line connecting the two on the diagram. Since the spectrum locus is everywhere either linear or convex, such lines always lie on or within this locus and represent all possible optical mixtures of all possible stimuli for each point. The diagram thus gives a tremendous simplification for this case. Three-part mixtures can be predicted by determining the point corresponding to any two of them, then this point to the third, and so on for more than three if desired. We might note that this property of the diagram is due to the definitions of x and y and also holds for any "linear transformation" of the diagram. The actual calculation of the mixture point is outside our present scope but basically amounts to the simple addition of the X's, Y's, and Z's in the required proportions and recalculation of x and y.

This straight-line rule makes it possible to visualize the vast number of energy distributions involved in a metameric set. Given a point on the diagram, any straight line passing through this point in any direction indicates a whole series of *sets* of potential metamers on each side of the point, any one of which can theoretically be mixed with any one on the opposite side to match any of those represented by the given point. The fact that this rule has been found to hold true in practice (for a single observer) indicates a remarkably stable property of the visual mechanism that contrasts vividly with the extreme mobility of the relation between the stimulus and what we actually see, a matter we shall discuss later.

Any three points that are not colinear determine a triangular area that includes all chromaticities that can be matched by real mixtures of the three. The largest possible such triangle obviously does not include all the points, thus illustrating the need for the imaginary primaries of the system. It is also apparent that a four-sided figure could include more points, and so on. If any of these areas include the point being taken as achromatic, then appropriate amounts of the primaries will match this point and can produce stimuli for any closed locus around it, that is, can produce stimuli seen as of all possible hues.

This raises the question of the meaning of the term "achromatic point" and the general subject of stimuli seen as colorless. Unfortunately, while we can usually justify the assignment for a particular purpose of a particular point as the achromatic point, perceptually it is completely arbitrary. There is *no* area within which the point *must* fall.

When we are using the diagram to consider the chromaticities of colored objects all illuminated by light of a single energy distribution, and this illuminant appears colorless under the conditions, then it is apparent that its chromaticity is the achromatic point. The phenomenon of chromatic adaptation, which we shall discuss in a later chapter, is responsible for there being a very considerable area of the diagram in the vicinity of the E point that represent stimuli that can appear to be without hue. This area appears to be roughly sausage-shaped, including the blackbody locus, from about 10,000 to around 4000 K and extending beyond both sides of it for a considerable distance, the distance being shorter across than along it. The subject is complicated by the fact that an illumination stimulus is often *accepted* as colorless when introspection would indicate that it is actually being seen as having a hue. The boundaries for such stimuli are thus very indefinite and much affected by circumstances. Nevertheless, for most purposes almost any point in the vicinity of the E point can be taken quite arbitrarily as achromatic. For practical purposes, however, as in the case of illuminated objects, it is necessary to specify not only such a chromaticity but also its actual spectral energy distribution. For this reason the CIE established three such "sources" in 1931, sources A, B, and C. They have recently added a fourth, D_{65}. These are all indicated in Figure 4-2, but they are defined as standards on the basis of their energy distributions. Source A represents incandescent lamplight at a color temperature of 2854 K; B, direct sunlight at a correlated color temperature (see Wyszecki 1967) of 4870 K; C, light from an overcast sky with a correlated color temperature of 6770 K; D_{65} is a simulated daylight distribution with a color temperature of 6500 K intended for use as a source when the short wavelengths are of particular importance, as in fluorescent "brighteners." Most of the material published to date that uses the CIE system has been on the basis of Illuminant (Ill.) C as the achromatic point. Use of this point facilitates perceptual interpretation of the diagram also, since it corresponds roughly to average daylight conditions. Actually, for many purposes the E point itself represents a satisfactory compromise between daylight and artificial light and thus permits comparison of direct spectrophotometric data without reference to an illuminant.

Given a fixed achromatic point, any chromaticity may be specified by its position on the line passing through it connecting the achromatic point with the spectrum locus. The wavelength of intercept is the "dominant" wavelength, so called because the color itself can be matched by mixing this wavelength with the stimulus representing the achromatic point. The position along the line, in the CIE system, is indicated by the ratio of its *distance on the diagram* from the achromatic point to the total distance from achromatic point to spectrum locus. It was designated by the CIE as "excitation purity" (p_e) and, of course, varies from 0 at the achromatic point to 1.0 at the spectrum locus. Unfortunately, although convenient for the designation of a position on the standardized CIE diagram, excitation purity has little relation to the appearance of the stimulus under any conditions except, perhaps, its appearance relative to that of other stimuli of the same dominant wavelength.

If the dominant wavelength line is extended through the achromatic point to the spectrum locus at the other end, its intercept gives the complementary dominant wavelength for the color. Stimuli from any point on this line can be mixed with the color to match the achromatic point. Note that neither dominant wavelength (λ_d) nor the complementary (λ_{dc}) have any meaning unless the achromatic point is specified.

When either the dominant wavelength line or its extension to the complementary intercepts the spectrum locus on the line connecting the ends of the spectrum locus, two minor problems arise. First, since positions on this connecting line represent mixtures, an awkward notation is avoided by referring to them by their complementary dominant wavelengths (λ_c). As we have just noted, this means that the achromatic point *must* be stated. A better system, which would be more readily interpreted, would be one based on the ratio itself; but no such notation has been developed. The second problem pertains to the purities of colors in this region. Since all points on the line (we need a name for it) represent mixtures of two stimuli, each with purities of 1, they all represent the highest purities that can be obtained physically and so represent a purity of 1 in exactly the same sense that the spectrum locus itself does; p_e can thus be calculated in the same manner, and only the notation need cause confusion.

As we noted in the previous chapter, a mixture of monochromatic and achromatic light can also be defined by its *colorimetric* purity which is the fraction of the luminance of the monochromatic component present in the mixture, thus also varying from 0 to 1. Because in the CIE system all luminance is assigned to the $\bar{y}$ function, excitation purity does not have a simple relation to colorimetric purity. Although they both necessarily increase continuously from 0 to 1, their relation depends rather heavily on both the dominant wavelength and the purity level. Unfortunately for the subject of color perception, it is colorimetric purity that, for a given dominant wavelength, most closely correlates with a single perceptual variable. (I shall later define "saturation" as "*apparent* colorimetric purity.") Points on the CIE chromaticity diagram can thus be extremely misleading in this respect. A diagram showing the loci of the chromaticities with constant colorimetric purity is given in Figure 4-3, for Ill. C as the achromatic point. The distortion is, of course, greatest at

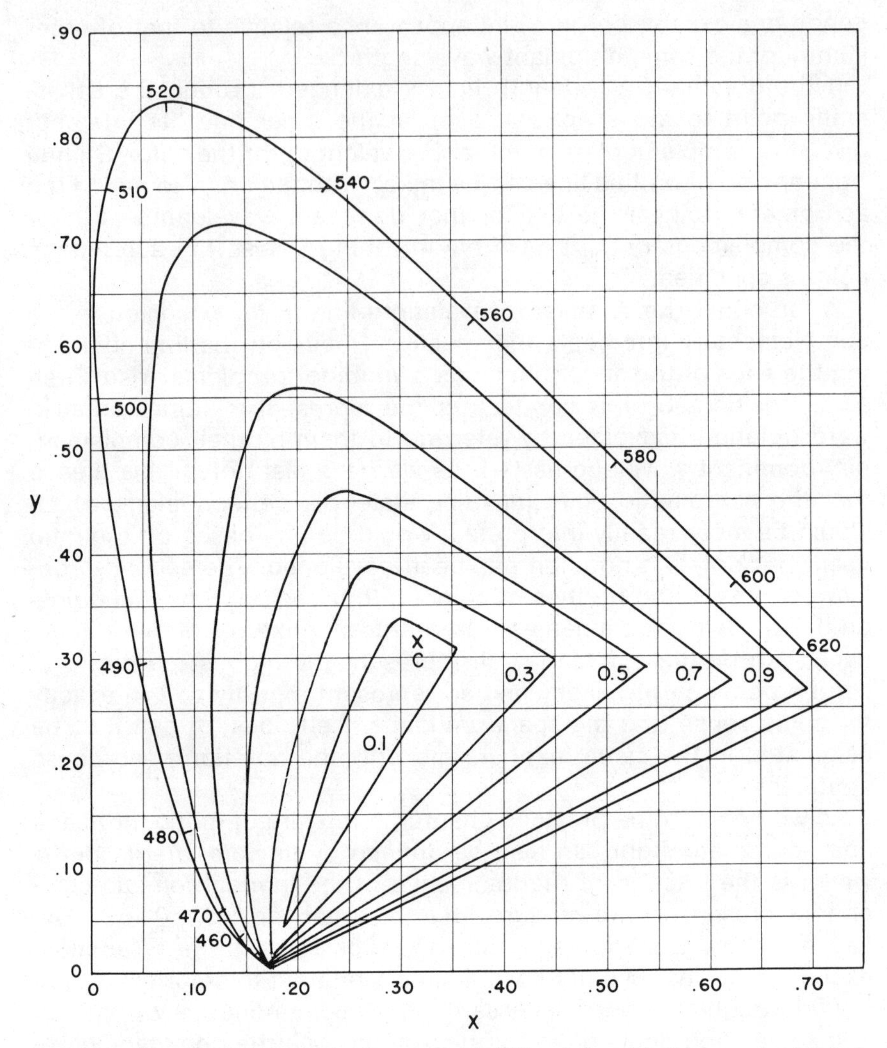

Fig. 4-3. CIE 1931 chromaticity diagram with "C" point and loci of constant colorimetric purity (p_c).

short wavelengths but is here very large; for example, for Ill. C, for λ_d at 450 when $p_e = 0.64$, $p_c = 0.09$; and the discrepancy is even greater at shorter wavelengths.

Only part of the diagram's distortion from the perceptual point of view is due to the artificial nature of the primaries. A considerable insight into the factors involved is afforded by a concept introduced by MacAdam (1938), which he called the "moment" of the chromaticities. Technically, it is the "mass" $(X+Y+Z)$ multiplied by the distance on the diagram to the achromatic point. It represents a combination of two visual characteristics of the different wavelengths, neither of which is obvious from the diagram itself. The first is the relative contribution of each to luminance as shown by the luminous efficiency curve. The second has to do with the colorfulness of each as evidenced by its effect in mixtures, particularly in the relative amounts of each needed to neutralize its complementary. This latter characteristic will be of considerable concern to us in later chapters; it is related to what I will call the "chromatic strength" of the colors. The loci of the chromaticities with constant moments are shown in Figure 4-4, taken from MacAdam's paper. The number associated with each locus is the moment per millilambert with respect to Ill. C. Colorimetric purity for any point on the diagram (for Ill. C) is the *ratio* of its moment to that of its corresponding dominant wavelength. Again, for any given point, the relative amount (millilamberts) required to neutralize (match Ill. C) any of the infinite number of its complementary chromaticities falling on the line from the point extended through the C point is given by the inverse ratios of their moments. In other words, the moments are direct indications of their relative strengths per millilambert in complementary mixtures *only.* This simple relation does not hold for the additive mixtures of stimuli that are not complementary. The distribution of these loci, nonetheless, does give a rather clear picture of what might be called the "relative importance" of the various areas of the diagram.

One major use of colorimetry, especially in industry, is the specification of acceptable tolerances in the colors of objects intended to be recognizably alike. It is obvious that such tolerances can be indicated for any point on the diagram, at least for chromaticity, by simply drawing a boundary around it, each point of the boundary representing the limit in that direction. It is equally apparent that the distances from the point to this boundary vary in significance both around the point and for points in different parts of the diagram. For

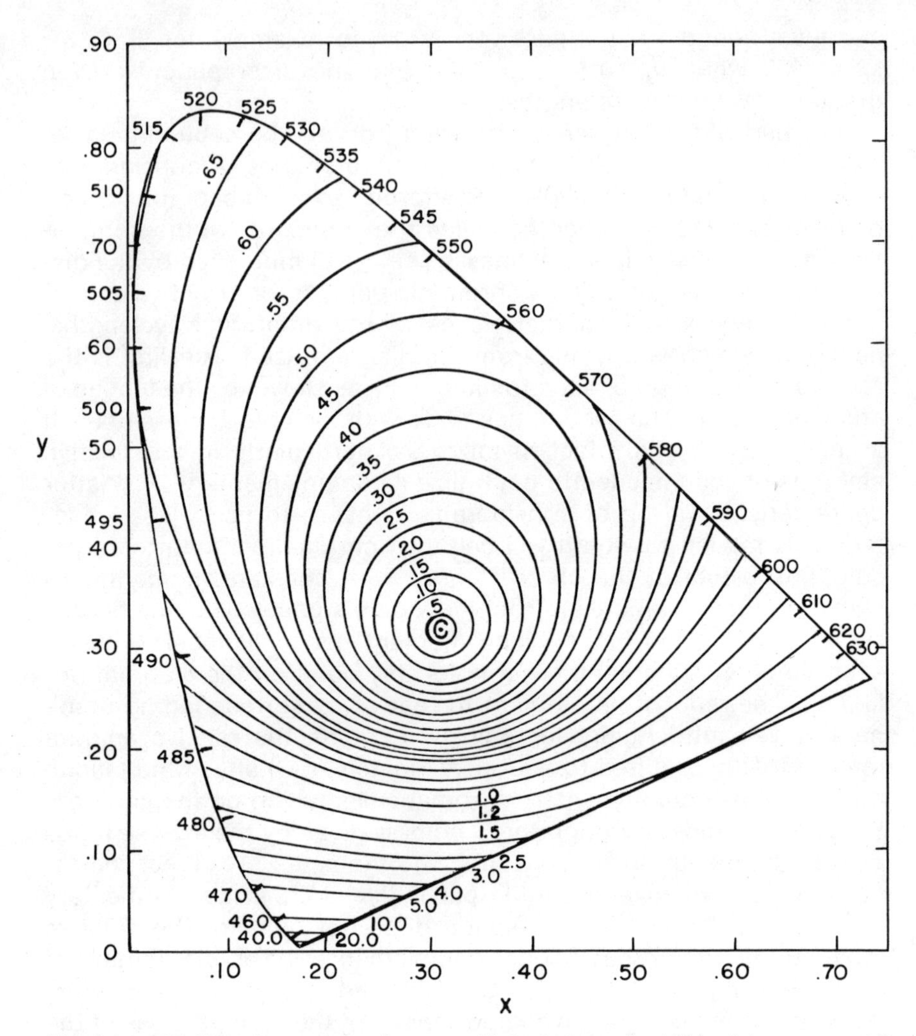

Fig. 4-4. CIE 1931 chromaticity diagram with loci of constant "MacAdam moment." (From *J. Opt. Soc.* **28,** 103, 1938.)

this reason there have been many attempts in recent years to find a metric in which equal distances have equal significance for all parts of color space. The subject is outside the scope of the present book but raises a question that we need to consider in some detail. Up to this point the whole CIE system is based on the facts of metamerism; the psychophysical conditions that must be met for two stimuli to *match* for the Standard Observer. The moment we start talking about the visual *difference* represented by two points on the diagram we have introduced a whole new subject. There is no *a priori* reason to believe that the variables appropriate to the calculation of a match are also appropriate to the calculation of this difference, although any experimentally determined facts could presumably be expressed by means of them if no *new* variable was uncovered.

One experimental approach to this problem is to assume that just perceptible differences are equivalent for all variables and significant in terms of their multiples. One of the well known studies of this nature that was made by MacAdam (1942) for chromaticity variables only, produced the ellipses shown in Figure 4-5, each of which represents the locus of points having 10 times the discriminable difference from the indicated point. Other studies have been made by other observers, and, of course, a complete study also involves luminance, but the figure serves to illustrate the problem. For our purposes we can simply say that no metric has yet been found that will convert any sets of such data to the required circles or spheres; no simple relations have been found. A tentative approximation has been put forward by the CIE as the 1964 Uniform Chromaticity System (UCS). The question that concerns us is whether the apparent absence of such a simple relation is a characteristic of the visual mechanism or whether, in fact, there is a new variable involved that is not apparent in the matching situation. Discussion of the matter will have to be postponed until we have systematically considered the perceptual variables. In this study, however, we will find that the simultaneous viewing of two nonmetameric stimuli *does* introduce a perceptual variable not present when viewing any single stimulus or single matched pair of stimuli. The second alternative thus seems the more likely one.

None of this, however, affects the fact that the 1931 CIE Standard Observer gives a unique specification for all possible visual stimuli nor that the CIE chromaticity diagram is an extremely useful tool in spite of its nonuniformity. So I want to conclude this brief sketch with a little further discussion of the meaning of the specification and a few notes on approximate relations indicated by the diagram. I regret that the brevity of my treatment gives no indication of the brilliance of the work that has been done in this field nor the size of the literature. Reference must be made to books such as those by Wright (1947, 1958), Judd (1952), and Wyszecki and Stiles (1967) for details.

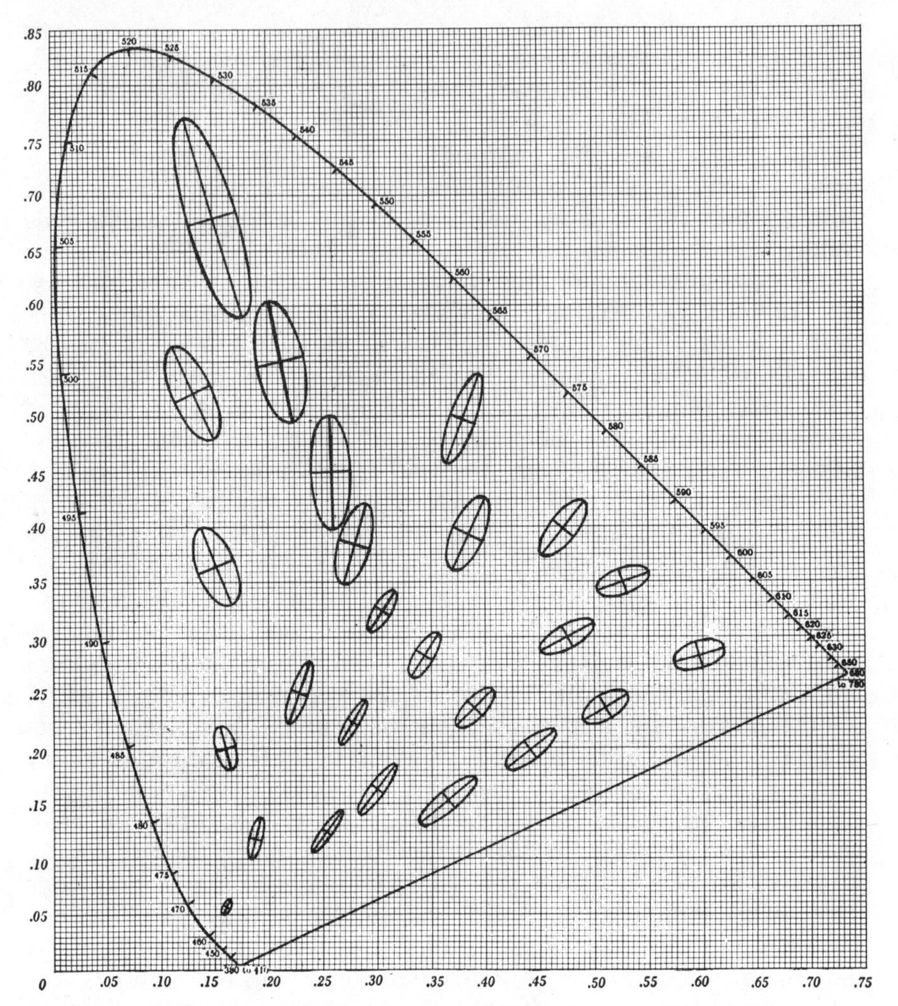

Fig. 4-5. CIE 1931 chromaticity diagram with "MacAdam ellipses." (From *J. Opt. Soc.* **32,** 247, 1942.)

FOVEAL VIEWING

The 1931 CIE Standard Observer is based on data obtained in a 2° field and accepted for use out to centrally fixated 4° fields. It is also restricted to photopic levels, which is assumed to be equivalent to stating that it is restricted to the fovea. What, roughly, does this restriction imply in terms of the viewing of stimulus areas in everyday life? The fovea of the eye is the area from which the greater part of our vision is derived. It is the seat of the most acute perception of form as well as color. When we want to see something clearly we "look directly at" it, which means that we cause its image to fall on the fovea. It is instructive to consider what is included in a 2° field and how this relates to everyday experience.

An angle of 2° at the eye corresponds closely to a circular area in the field of view whose diameter is equal to 3½% of the distance, and a true "rule of thumb" can be given for indicating it directly. For almost anyone, if he holds one arm out directly in front of himself, with the thumb extended vertically, the image of the outer segment of the thumb will subtend about this angle or will, in any case, fall well within the fovea. With this as a guide, the process of seeing can become apparent quickly and often surprisingly. Broadly speaking, no other part of the eye is ever used to see whatever is of interest at the moment, and often it is only the center part of the fovea that is used. Consider a person 6 ft tall walking along the other side of the street, say 30 ft away. At this distance 2° is about 1 ft. If you are interested in his clothes you look him all over, meaning that you successively make different parts of the image fall on the fovea. At 10 ft the area is about 4 in. across, and yet if you are interested in an object of about this size you still look at first one part of it and then another. At reading distance, say 20 in., the width has reduced to about ¾ in., smaller than some of the words; and reading becomes a series of quick successive fixations along the line with occasional returns to a single word for verification. It is thus safe to assume that any stimulus with which we are critically concerned will be seen by foveal vision at photopic levels. We shall find, however, that the area that surrounds a stimulus can play a large part in what we actually see, and we also know that the characteristics of the eye change fairly rapidly and nonuniformly outside the fovea. For both these reasons we are interested in the colorimetry of fields larger than 4°. The required studies have been made for 10° fields, largely by Stiles (see Wyszecki 1967), and a 1964 CIE 10° Supplementary Observer has been adopted which can be used for such purposes. The differences are not large enough to concern us at the moment, but we know this only because the data are available. In like manner the CIE has adopted data for the parafoveal wavelength-relative efficiency function for scotopic intensity levels. This again is beyond our present scope but it does indicate quantitatively how parafoveal metameric pairs become dissimilar at these levels. The progressive stages of this change over the mesopic range have yet to be quantified.

THE CHROMATICITY DIAGRAM

We turn now to the wholly different subject of the various ways in which the chromaticity diagram based on the 1931 CIE Observer can help us visualize the general approximate relations between stimuli and the perceptions they invoke. I introduce this here rather than later because thinking in terms of the chromaticity diagram can be tremendously helpful even in cases where it does not offer a direct solution to a problem.

We shall see, in the next chapter, that there is a fairly direct relation between the chromaticity of a simple stimulus and the color seen. Colored signal lights are a typical example, and for this case Kelly (1943) has published a CIE diagram in which the general perceived color is shown for the various areas of the diagram. This is reproduced in Figure 4-6; it provides an excellent way of visualizing the "meaning" of the different areas of the diagram, but it must be remembered that the names apply only to hue (except for pink) since all the colors weaken continuously to achromatic at the center.

While it is true, as we shall see, that stimuli represented by almost any point on the diagram can be made to appear of almost any hue (and this is particularly true of those near the center), it is also true that, by and large, stimuli will tend to have the hues indicated. This is true not only for the light sources for which the diagram stands but also for colored objects seen in everyday circumstances. Thus the chromaticity point corresponding to the absolute spectral reflectance curve of a surface color, while strictly valid only for Ill. E, will indicate the approximate color of that surface as usually seen. If it does not, in a particular case, then either the illuminant or the reflectance characteristics are quite unusual. Among occasionally notable exceptions are some commercial fluorescent lights, highway sodium and mercury lights, and some flower colors. In a similar manner, the statement that two colors are complementary, or the expression "complementary colors," means simply that the colors are roughly opposite on this diagram across a point somewhere near the center.

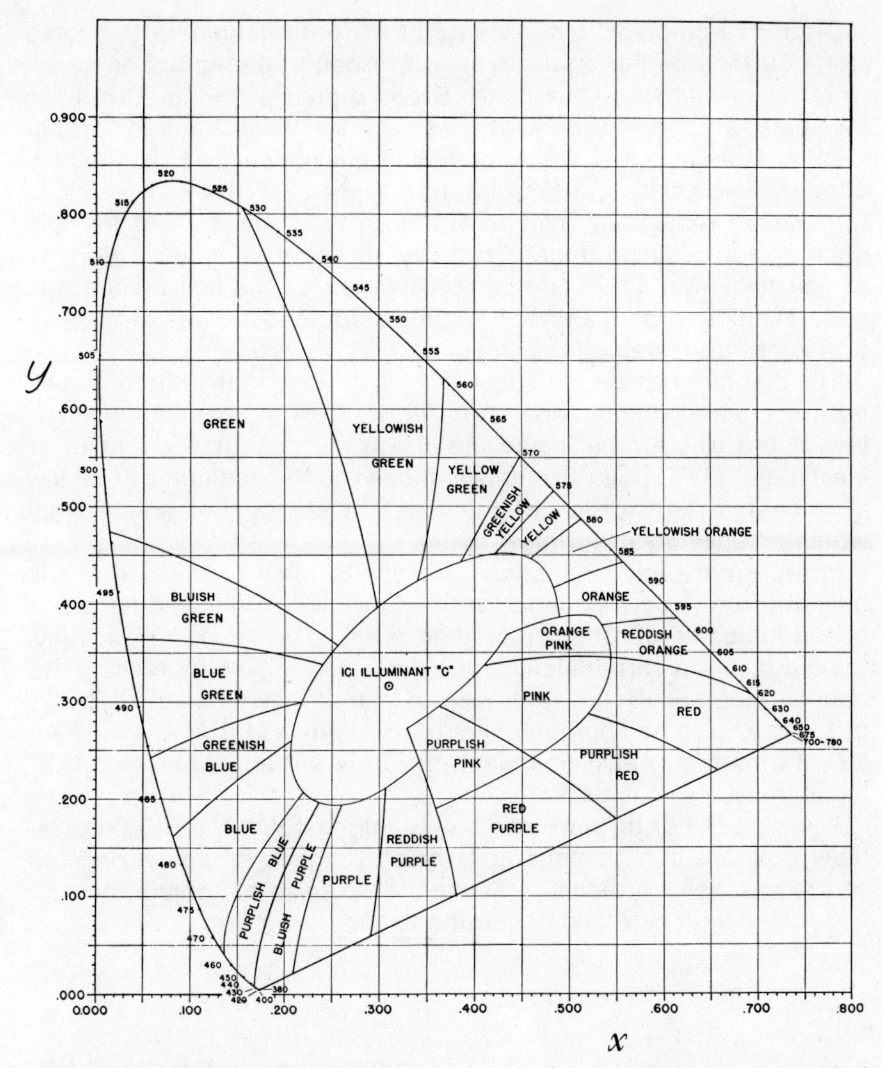

Fig. 4-6. CIE 1931 chromaticity diagram showing color designations for lights, by K. L. Kelly. (From *J. Opt. Soc.* **33,** 627, 1943).

Of course colored lights, when mixed (both illuminate the same white surface or the equivalent), will produce the colors lying between their *names* on the chart. Even colored pigments, if not too far apart, will tend to follow this same rule. While the color of pigment mixtures such as watercolors and oils cannot be predicted accurately except by considering the mixture of their absorptions rather than reflections, two colorants in the same part of the diagram will usually produce a color lying between them. Colorants whose reflection colors are complementary on the chart will not in general mix to produce an achromatic color, and the hue of the mixture is not predictable from the chart.

The diagram can even be used to give a rough indication of the effect of colored illumination if the purity of the illuminant color is low. If the whole area inside the spectrum locus is thought of as elastic and the C point is mentally moved to the position of the new illuminant, the distortion of the diagram will be not unlike what would actually occur for most colors.

In fact, more misconceptions are apt to arise in the use of the diagram from failure to consider the possible luminance differences of the indicated chromaticities than their hues. Thus, for example, the locus of the chromaticities of a dye solution with increasing dye concentration will follow a curved path that may approach a purity of 1. It may also be approaching a transmittance of 0. Two nearly adjacent points may also represent very different appearances if their luminances are much different.

The purpose of these remarks is to suggest that the CIE chromaticity diagram is a tremendously helpful means for *visualizing the nature* of a color problem. Any use of it to *solve* a problem must be conducted with care and according to the rules.

Five

ISOLATED STIMULI: AFTERIMAGES

Up to this point we have been concerned primarily with the physics of the stimulus and the way in which stimuli can be classified, based on certain well established facts concerning the eye as a receptor. We have seen that all possible color stimuli can be arranged in orderly fashion in three dimensions and that this arrangement can be achieved largely without reference to the colors actually seen. We need now to consider in some detail the nature of the relations between the psychophysical variables by which the stimuli have been arranged and the colors seen at various points of the arrangement. Because of the relative inflexibility of the rules by which stimuli can be assigned to metameric groups, there is always a tendency at this point to assume that this inflexibility must somehow extend to the perceived colors. In one sense it does: if all the external light stimuli at any one moment can be specified and the observer can be considered to be in a standard state, then the stimulus does specify a unique range of responses. In the general case, however, there are far more variables external to the stimulus that affect the perceived color than there are in the direct stimulus itself. For this reason it is more productive to consider the perceptual variables by themselves and to develop a descriptive terminology for them that is independent of the stimuli. The effects of the stimulus variables, both direct and external, can thus be described in terms of this fixed perceptual framework and, for typical *situations,* a direct relationship established.

The simplest case—one can almost say the only simple case—in which a color is seen occurs when a single, uniform stimulus is the only one sending light to the eye. Under these conditions there is a fairly well established relationship between the psychophysical and the perceptual variables. We can thus use this case to start our arrangement of perceived colors in a space of their own. We shall consider this in some detail in the present chapter, but we must keep in mind that this situation does not introduce *all* of the perceptual color variables. The others and the situations that produce them form the subject matter of the second section of the book. The assumption that this simple case does produce all the variables has been the cause of much ambiguity in the literature of color, particularly in popular presentations of the subject, and it is the resolution of this ambiguity that is the main purpose of this book. On this we can make a good start by noting at once that the fact that we shall find more than three *perceptual* variables does not affect in any way the fact that all single color stimuli may be characterized by only three *psychophysical* ones. In most cases we deal with more than one stimulus, each characterized by three, and it is not surprising to find perceptual interactions among them.

We can start, by considering the color perceptions producible by a single light stimulus, "isolated," in the sense that it is the only stimulus in the field of view, and sending constant and identical light to the eye from every point of its area. The area can be specified by the angle it subtends at the eye and the quality by its chromaticity and luminance or variables derived from them. For present purposes I shall use: visual angle, wavelength for monochromatic light, dominant wavelength and colorimetric purity (both to III. C) for nonmonochromatic light, and luminance. For simplicity I shall first consider the relations quite generally without regard to visual angle, introducing its special effects afterward.

One of the perceptions produced by an isolated, monochromatic stimulus is hue, and this hue is heavily dependent on the wavelength. This dependence is very far from linear throughout the spectrum, and there is no *a priori* reason to expect that wavelength as such is the most appropriate variable to use. Wavelength came into use at a time when it and the velocity of light were the easiest variables to measure, and so wavelength has become established as the customary variable. Actually both are constant only in a vacuum, the invariant being the frequency, which is independent of the medium. In recent years there has also developed, largely for nonvisible radiation, the use of the wave number, defined as the number of cycles per centimeter. Thus the question often arises as to whether one of these other variables would be preferable to wavelength. Actually, since they are all simply related, it makes little difference which one is used. The basic psychophysical function involved is that of wavelength discrimination at constant luminance, and this is not simply related to any of them.

A linear scale of hues can be produced only by empirical studies and the results then related to one of these variables. I shall use wavelength throughout in order to relate better to the literature even though it seems to me more likely that one of the others would be preferable if our subject were color vision theory. In any case, we also have to contend with the fact that not all the hues are producible by monochromatic light of a single wavelength so that we must be somewhat arbitrary in defining the stimuli that produce these "extraspectral" hues.

HUE

The necessary basic requirements of a hue scale were discovered and stated by Newton. He found that all possible hues could be produced by monochromatic light either alone or in mixtures of pairs, and that the mixtures of pairs always produced either a hue intermediate between the two in the series or an achromatic color. Hues thus form a closed circuit, of which only part is producible by single, monochromatic stimuli. The names of the hues in this circuit have been confused ever since Newton. He named seven hues *in the spectrum* and showed a closed color circle divided into these seven, which he called red, orange, yellow, green, blue, indigo, and violet. In his text, however (see MacAdam 1970), he notes that if the color falls near the dividing line between violet and red, then the color "is not any of the prismatic colors but a purple. . . ." In fact, therefore, he names eight colors for the *complete* circuit. There has been much discussion over what Newton meant by "indigo". His use makes little sense as long as "blue" is considered to have its present-day meaning. If, however, his "blue" is interpreted as our "cyan," and "indigo" is taken as our "blue," then his eight colors line up fairly well with the divisions I shall adopt. The assumption is not unreasonable if "blue" is defined as the color of the sky. Since hue is one of the basic color perceptions it is important that we consider this in some detail.

Since hues form a continuous closed circuit we could start at any point, but it is convenient to start with monochromatic light from the long-wavelength end of the spectrum. If we have such a stimulus, say 700 nm, seen as an isolated stimulus, and ask several observers to name the color they see, we can get a variety of enlightening responses. All (visually) normal observers will agree that the color is primarily red, but some will insist it is *the* red, and others will insist it is a distinctly yellowish-red. The difference is undoubtedly due to experience with colors, those familiar with the spectrum calling it red and those familiar with colorants calling it yellowish-red. Fortunately there exists a criterion for what *should* be called red from a purely logical standpoint, and we will return to the matter presently. If now we decrease the wavelength of the stimulus continuously, keeping luminance roughly constant, the yellowish character of the color increases steadily. At some point we arrive at a color the observer would call orange. This is a very distinctive color but there is little question that it partakes of the nature of both yellow and red in about equal parts. This is an important point of view though some observers will argue against it, saying that orange is a separate hue. The fact is that if the wavelength can be varied continuously through this point, the color will become redder for longer wavelengths and yellower for shorter *without* wholly losing the character of either. It can thus be considered as a mixture of the two hues in spite of its characteristic appearance. Continuing toward shorter wavelengths, a color is reached that no longer has any trace of red and that the observer would agree is pure yellow. This color has an affinity toward red but is unmistakably different. The disappearance of the red takes place quite abruptly in terms of wavelength, and as we pass through the yellow point there is an equally abrupt appearance of a greenish component in the color. Yellow is thus a simple hue in the sense that it is not seen as a mixture. Hering (see Hering 1905–1920) called it a "primary" hue but I will follow more recent practice in calling it a "unitary" hue.

As we proceed toward shorter wavelengths, the greenness of the hue increases, and a point can be found where green and yellow appear about equal. This point is poorly defined and we have no common name for it, although it is often called "olive" green. Newton omits this color. By continued shortening of the wavelength, we arrive at unitary green, beyond which blue appears in place of the yellow. For many observers, including myself, neither this green nor the equal part yellow-green have the distinctive character of yellow and orange; and the unitary green, seen by itself, looks distinctly yellowish. There is no doubt, though, that a wavelength can be found that for a particular observer looks yellower if lengthened and bluer if shortened. We can thus define unitary green as the color between these two. Continuing, we find color with apparently equal parts of blue and green, to me nearly as distinctive a color as orange. This color is often intended by the words "turquoise" and "cyan," but again, there is no commonly accepted name. Beyond this is found unitary blue and this, like yellow, is unmistakable, changing to reddish-blue beyond, but showing no tendency itself toward either green or red. As in the case of yellow, the transition is an abrupt one in terms of wavelength. At the shortest wavelengths red is seen as a component, and the color is called "violet." This color tends to be somewhat ambiguous visually in that, under some conditions of adaptation, the red component tends to dissociate itself spatially from the blue. Monochromatic stimuli in this region are both difficult to obtain physically at high intensity and purity and to look at when obtained, easily producing fear of damaging the eyes. There is also at the shortest wavelengths an increasing fluorescence of the ocular media that further confuses matters. Nevertheless, it appears established that if monochromatic light from the long-wavelength end of the spectrum, say 700 nm, is now added in increasing amounts the transition is continuous into the reddish blues. This region needs much more study from the standpoint of perception. As further long-wavelength light is added, we again arrive at a poorly defined point where red and blue are about equal in importance. This whole region of colors is called "purple," but there seems to be no agreement on which is *the* purple. As the ratio of long to short wavelengths is further increased, a point is reached where blue disappears and yellow takes its place. This occurs distinctly *before* the mixture is 100% long-wavelength light. This color is, again, and almost by definition, unitary red. Just as unitary green has to me a strong yellowish appearance, so also this color has to me a strong bluish appearance when seen by itself, and I would tend to name it "pink" rather than red. The yellowness increases, however, until 100% of the long-wavelength component is reached, and the series is then continuous with that obtained by decreasing this *wavelength,* which is where we started. The circuit of hues can thus be considered as completely continuous and as having four unique, unitary hues, visible in mixtures only *between* them in the circuit. Of these unitary hues, two, the blue and yellow, are totally unlike, and the green and unitary red have, to me at least, strong affinities for yellow and blue, respectively. Of the mixtures, orange and cyan are distinctive, and yellow-green and red-blue are considerably less so. Of course, these are personal, subjective judgments; no other attitude except consensus is possible in perception. The literature does not yet offer a consensus on this subject.

The preceding paragraph describes a field obviously fertile for controversy on how the circuit of hues "should" be divided, and it has been quite thoroughly cultivated. For present purposes I shall accept the four unitary hues as facts, although I do not find them much more distinctive *as colors* than orange or cyan, nor, for that matter, than the pink and brown we will encounter later. As a matter of interest, this scheme lines up completely with Newton's if a term is added for yellow-green and his violet, like ours, is considered part of the purples. This gives four unitary hues and four intermediates.

These unitary hues are unique perceptions that we can use for reference points and for descriptions, but they must not be thought of as unique responses to particular stimuli. If hue only is considered, they can each be produced by an infinity of mixtures of stimuli as well as by changes in the adaptation state unrelated to the stimulus *per se.* Nevertheless it would be a considerable contribution to the subject if, for the case of an isolated stimulus with a given visual angle and luminance, a hue-wavelength conversion table could be accepted for general use. The function is so irregular that any set of data that describes a visual response in terms of wavelength is bound to be misleading. On the other hand, the difficulties involved in reaching such an agreement are rather formidable. They are, however, quite instructive and probably no more severe than those originally encountered in accepting a norm for the luminous efficiency curve. Thus they involve wide variations between normal observers, some but not many semantic problems, and distinctly arbitrary decisions as to the manner of handling the lack of hue change at the long-wavelength end of the spectrum and the complex nature of violet at short wavelengths, both in relation to the way in which the extra-spectral stimuli are merged with the spectral ones.

Agreement on such a conversion table, quite aside from its future usefulness, would be important in resolving two further problems. We have hue-wavelength scales, particularly in the Munsell system, that have been equally spaced with great care. They apply, however, only to the case of object colors *seen against a background;* it is not known whether they also apply to isolated stimuli in detail. We shall find that the background is the deciding factor in hue relationships. We do not know, except broadly, whether *any* background produces the same relations as those for isolated stimuli and so have no norm against which to consider the effect of the *presence* of background. Conversely, if we take as the norm selective samples against a nonselective surround, arbitrarily lighted by, say, Ill. C to which the eye is assumed to be adapted, then we do not now know the effect of adaptation to the stimulus *itself* that we must also assume to occur with isolated stimuli.

The major difficulty in establishing the latter relationship lies in the fact that *no* direct stimulus comparison can be made; that is, no two stimuli may be seen by the same eye at the same time that the judgment is made. There thus appear to be only three main approaches to the problem: successive comparisons, verbalized distinctions, and inter-eye comparisons. Each of these approaches has its own difficulties, but the attempt at quantification of the situations would itself add greatly to our knowledge. The first approach involves short time color memory and the uncertainties of the time course of adaptation. The second involves the whole problem of the semantics of color names and hence requires a large number of observers and a limited number of points but should not prove to be either impossible or ambiguous. The third method involves the well established fact that there is little interaction between the adaptation states of the two eyes; and it appears to be the most promising approach, not only to this problem but to many others in color perception. It is so promising in fact that it is worth digressing to consider it for reference later; for some problems it seems to offer the only hope of a solution.

The fact that the two eyes can be and often are in different sensitivity states simultaneously was known to many of the earliest workers; it is a fact easily demonstrated by covering up or shining a light into one eye. It has also been known that under some circumstances there is fusion of colors presented separately to each eye. This latter effect has been found to be the exception rather than the rule; in most cases what is seen with distinctly different stimuli is an alternation of the two, the so-called "retinal rivalry." The pioneering work of Wright (Wright 1947) showed that the responses of the eyes are so completely independent that thoroughgoing studies of the effects of adaptation may be made by holding one eye constant and varying the other, even over extreme ranges. We shall see presently that the greatest gap in our knowledge of color perception involves the hue that is seen from a given stimulus when the conditions under which it is seen are altered. This possibility of "inter-eye comparison," as I shall call it, seems to me to offer the only accurate means available to supply this much needed information. Considerable work has been done in this field in recent years, some of which we will note in passing, but very little of it has dealt specifically with hue. The possibility of presenting different *situations* to each eye and directly comparing the colors seen, *without having them affect each other,* offers hope of eventual development of quantitative relationships between the psychophysics of the stimuli and the colors seen. I shall have frequent occasion to refer to the possibility.

Associated with the problem of hue-wavelength relations for isolated stimuli is the effect of the luminance of the stimulus at a given wavelength. It is usually stated in the literature (Troland 1929, for example) that there are decided hue shifts characteristically different for different wavelengths and that there are four "invariant hues" that do not show such a shift. The results usually quoted to support these statements are those obtained by Purdy in Troland's laboratory; it is not at all certain that they hold for truly isolated stimuli, nor do we know the extent to which they depend on visual angle or duration of viewing. What is definitely established is that when *two* monochromatic stimuli are present at different luminances but the same wavelength, then there are very definite and quite large hue shifts as above. It is also often stated that at "very high" intensities all stimuli tend to shift toward either yellow or blue, but it is not known whether this is simply a matter of "overloading" the visual mechanism. It seems more likely that if adaptation were properly controlled and account taken of the effect of visual angle, the hue-wavelength relation would be found to be more or less independent of luminance but very strongly dependent on momentary adaptation state. Thus, for example, hue changes do *not* occur over the very great intensity changes which take place in normal daylight. In any case our present purpose has been served by noting the nature of the relationship.

ACHROMATIC STIMULI

Before discussing the perception of isolated mixtures of monochromatic stimuli, it is first necessary to consider more closely the nature of isolated stimuli seen as achromatic. As we have noted earlier, there are a very large number of sets of metameric stimuli that can produce this perception and a considerably smaller number that usually do. Whether there is a single chromaticity for a given observer that always produces this response instantly after total dark adaptation or whether many also fulfill this criterion still seems to be somewhat unsettled. The former would suggest that there is a fixed chromatic "resting state" of the eye, and the latter would suggest that adaptation to the stimulus is coincident with the perception of the stimulus. The alternative, that there is a region of stimuli in which hue cannot be perceived when isolated, is negated by the fact that hue discrimination of stimuli *seen in pairs* is not abruptly different in this chromaticity region.

A careful study of a series of stimuli falling along the blackbody locus was made by Hurvich and Jameson (1951). Their study included stimuli from around 3000 to 10,000 K, with angular sizes from 2.02 to 46.8° and durations of 1 and 5 sec presented after low level "neutral adaptation." While they were interested primarily in a "white" threshold that does not concern us here, their results do indicate that at low luminance levels (1 to 10 mL) there tends to be a quite narrow range of color temperatures that are seen as achromatic with 1-sec exposure. This range varied considerably among the three observers and also depended somewhat on field size. For two of the observers this point was between 5000 and 5500 K and for the third about 7500 K; thus, for all it was in the range from sunlight to daylight. Stimuli at higher color temperatures were seen as blue and those at lower as yellow. As luminance was increased, for all conditions, the range of stimuli seen as achromatic increased, covering a range from around 4000 to beyond 10,000 K by the time the luminance reached 100 mL. The fact that these results depend on the time of exposure was shown by the difference between the 1- and 5-sec results, the 5-sec exposure reaching a wide range of stimuli at much lower luminances. There is no question that this difference is due to adaptation time; they point out that the decision was based on the appearance at the *end* of the 5 sec. Presumably a similar effect of less noticeable magnitude occurred also for the 1-sec exposure. The increase in the range with luminance also implies that the more rapid adaptation to the stimulus due to higher luminance is the cause of the loss of hue perception over the wider range. The existence of an achromatic "point" at low intensities and short durations does suggest, however, that there is in fact a single color temperature that appears achromatic to the dark-adapted eye. This apparently occurs at around 5500 K, roughly the color temperature of sunlight; this result is confirmed by many statements of earlier observers. (The chromaticity of 5500 K is also close to that of the "equal energy" source E.) It is interesting to note that color temperatures below about 4000 K were all seen as yellow up to the highest luminances (c. 500 mL) used. Thus in their tests all customary incandescent sources would have produced yellow. These results confirm the facts that CIE Ills. C and D_{65}, as well as E, are well within the limits acceptable as achromatic, at least after brief adaptation, that there is a large range of chromaticities which meet the same requirement, and that there may be a more or less unique chromaticity that under some circumstances may serve as a sort of reference point in perception.

SATURATION

We need to consider next the appearance, as isolated stimuli, of mixtures of monochromatic light with a stimulus that would be seen as achromatic under the conditions. (Again I shall consider this stimulus to be Ill. C but from the above a somewhat lower color temperature might be more logical.) Suppose that we have an isolated stimulus in which the ratio of monochromatic to achromatic light can be changed without changing the total luminance and that this is well above the photopic threshold, say at 100 mL. As we increase the monochromatic component from 0 at the achromatic point toward 100%, a series of purities are generated that can be identified either as colorimetric purity (p_c), based on relative luminance, or excitation purity (p_e), based on the distances on the CIE diagram. (The limits are 0 and 1 for both.) For very low purities the stimulus is of course seen as achromatic for all wavelengths. For all wavelengths (including the extraspectral stimuli) there is a definite purity for each at which a characteristic hue appears. This is known in the literature as the "purity threshold" although, for reasons given later, I shall call it the "chromatic threshold." The purity at which this threshold occurs is heavily dependent on the wavelength and less so on the area and luminance. For the very short wavelengths it can occur at purities as low as 0.001 p_c and for wavelengths seen as yellow as high as 0.25 p_c. We shall see in later chapters that this threshold is quite important perceptually. At purities above this threshold, both the achromatic and the chromatic component are seen as a mixture, with the monochromatic component increasingly dominant as purity increases up to 1, beyond which it is not possible to go. This series ends abruptly at the spectrum locus. When the series is run the other way (for isolated stimuli), Wright (1935) found that the first perceptible purity step is more or less the same for all wavelengths. Further investigation of this important fact is needed.

The main perceptual variable in such a series is called "saturation," following Helmholtz. There has been a good deal of confusion in the literature as to the meaning of this term, in part because Helmholtz himself used it in several senses. I shall use the term throughout to mean "the perception of *apparent* colorimetric purity." For isolated stimuli it has the value 0 at the achromatic point and increases more or less linearly with purity up to a purity of 1, beyond which the stimulus cannot go. It would appear to follow from this definition that two colors having the same purity would be judged as having the same saturation, at least seen as isolated stimuli. This is distinctly contrary to the facts, but I shall have to postpone further discussion of it until Section II of the book. We should note here that the saturation produced by a color of a given dominant wavelength is not necessarily limited by the fact that the stimulus cannot exceed a purity of 1. Thus, if the monochromatic stimulus itself produces a partially achromatic response, as may be the case for monochromatic yellow, the *apparent* purity would be less than 1 at an actual purity of 1. Alternatively, it is equally possible that for some short wavelengths the achromatic component may cease to be visible before a purity of 1 is reached. The facts in regard to these possibilities are not at present known. For this reason saturation cannot be *identified* with purity, and perhaps a better way of thinking of it is as the *apparent* percentage or fraction of the monochromatic component.

The perception of saturation is entirely analogous to that of mixtures of the unitary hues, in which both components are visible in, say, orange, and an estimate of the percent of the red component may be made. In the same way, in a saturation series achromatic and pure hue are seen separately, and both may still be present when the stimulus reaches a purity of 1. Whether this is actually the case for yellow seen as an isolated stimulus cannot be decided without further study, probably by inter-eye comparisons. That the different monochromatic stimuli produce different saturations and that each saturation can be exceeded under the proper circumstances have been considered truisms since the time of Helmholtz, but we will find that the word "saturation" has not always been restricted to perceptions that include an achromatic component.

BRIGHTNESS

We have thus seen that for an isolated stimulus at a constant photopic level of luminance, dominant wavelength predominantly controls the perceptual variable hue and that a mixture of monochromatic and colorless light predominantly controls the variable saturation. From the previous discussion we know that these factors cover all the possible stimulus chromaticities. The only other psychophysical variable is luminance itself and we now need to consider this again, still restricting ourselves to isolated stimuli. We saw in the previous chapter that luminance predominantly controls the variable brightness. It is often stated that luminance also affects both hue and saturation for a given stimulus. That this is true for stimuli seen in many *situations* cannot be questioned; that it is true for isolated stimuli remains to be demonstrated. In any case, these effects may be neglected here. For a stimulus of any chromaticity, if the luminance is increased brightness increases, and, introspectively, there is an absolute brightness associated at least roughly with different luminance levels. Thus we speak of a given stimulus as appearing very bright, dim, dazzling, and so forth. In this sense, we can say that luminance controls brightness. Since luminance is a calculated variable, based on the luminous efficiency function which is itself based on equality-of-brightness data, we would expect to find that isolated monochromatic stimuli of different wavelengths but the same luminance would appear to be equally bright on an absolute, introspective, basis. This is found not to be true; a high purity blue stimulus appears much "brighter" than a corresponding stimulus of longer wavelength. We thus encounter for the first time the "other" perceptual variable that makes heterochromatic brightness matches at high purity ambiguous. I shall again delay consideration of it to the next section and note here only that the perceived differences due to this variable are very large and heavily dependent on wavelength; at 100 mL in a 10° isolated stimulus, for example, λ430 nm is dazzling and λ574 nm can be considered as only comfortably bright. It is because this "other variable" as seen in an isolated stimulus is very much like brightness that it has been customary to say that hue, saturation, and brightness are the *only* perceptual variables involved in such stimuli. This belief has been supported by the fact that only three psychophysical variables are involved and that luminance "works" in the calculation of matching stimuli that have widely different spectral distributions. We can anticipate our later conclusions by stating at this point that it is this other variable that the experts had to *learn to ignore* in order to make a luminance match, visually, between widely different wavelengths, and that the phase of the stimulus that controls it is inherent in the *color mixture* data rather than in the luminous efficiency curve. It is closely allied to MacAdam's "chromatic moment."

SIZE OF THE STIMULUS

We need to consider briefly the effect of the angular size of isolated stimuli on the perceptual variables. We need not go into this in great detail because our concern is more with everyday seeing than with the details of the mechanism, but size conveniently introduces the matter of adaptation and "afterimages" which is of great importance to the subject. Here again, we are dealing with subjective phenomena that have not yet been subjected to quantitative analysis; such analysis can probably be done only by inter-eye comparisons. Such reports as we do have are suspect because we do not know how to interpret a statement such as "the saturation decreases." With this in mind, however, we can rather quickly summarize the effects involved.

We noted earlier the loss of blue sensitivity (central tritanopia) for stimuli of very small angle (c. 10 in.) accurately fixated. This appears to have little effect on average vision but may well increase resolution of high contrast fine detail and perhaps plays a part in "vernier acuity." At larger angles, out to 4 or 5°, vision is largely foveal and is the type of seeing with which we are mostly concerned. Beyond this angle we encounter the "macula lutea" and the somewhat different color sensitivities of the outer retina.

For stimuli in the photopic range and larger than "points," we can make certain general statements. When first seen by a dark-adapted observer such a stimulus may appear quite bright, but this brightness decreases very rapidly (perhaps $1/5$ sec), and then may or may not continue to decrease slowly for some time. In general, both of these effects are greater for larger sizes and higher luminances. Coincident with this brightness change are color changes that are not well established. The effect is usually described as a loss of saturation but is probably more complex than this. Hue shifts are not usually reported but may occur. These perceived changes are usually attributed to the adaptation of the visual mechanism to the stimulus and involve a number of effects that we can consider separately.

Consider first a 10° circular field with moderate luminance, say 100 mL, uniform, with a sharp boundary, and first seen as of rather high saturation. If the center of this field is rigidly fixated and viewed without blinking (both difficult), there will be a gradual loss of both brightness and saturation over the whole area that can go far toward making the stimulus disappear. This progressive change can be interrupted at any point by either blinking or moving the eye quickly (changing the fixation point) from side to side. Eventually, after perhaps a minute or more of relaxed normal viewing of the center of the area, an equilibrium is reached at what I call the absolute brightness. Under these conditions, after fixed viewing for a few seconds, if the eye is moved to a new fixation point, say one-fifth of the diameter toward the right hand, fringes will be seen on each side of the stimulus. The fringe on the right will look dark, perhaps black, and the one on the left will look bright and of higher saturation, that is, will look more like the original stimulus. It is apparent that both these fringes are due to the fact that there has been a loss in sensitivity of the local area of the retina covered by the first image. The dark fringe is due to a decrease in the normal dark activity of the eye and the bright to the fact that a fresh, unaffected area is now receiving the image. I shall call this effect "local adaptation" of the retina; we need to consider its nature in some detail.

Note first that this is a dynamic equilibrium effect, subject to fairly rapid change. Closing the eyes for a moment, or even just blinking, starts the process again at a considerably higher sensitivity level. Closing the eyes does not eliminate perception of the response to the stimulus; a continuing, very unstable image remains. This image is presumably due to regenerative effects taking place during the "dark adaptation." Such phenomena can be quite complex, leading to what is known as the "flight of colors," but I consider them as outside our present study. I do want to consider the actual change that has taken place in the retinal sensitivities. This can be described best as the perceptions produced by the rapid substitution of a new stimulus in place of the one with which the eye is in equilibrium. We need consider only a few of the many possibilities. For convenience, think of these in terms of a smaller, say 4°, original stimulus.

AFTERIMAGES

Suppose that the quickly substituted second stimulus is much larger than the original stimulus, with the same luminance but the chromaticity of Ill. C. Except in the area of the original stimulus this will be seen, almost at once, as achromatic. Within the area of the original stimulus, the new stimulus will appear darker, of a hue roughly complementary to the original stimulus, and of roughly the same saturation. This is called the "afterimage" of the stimulus, as is the image seen on closing the eyes, but I shall restrict my use of the term to those seen due to a second stimulus. This image moves about if the direction of the eye is changed but retains the same visual angle and position with respect to the eye axis. Its boundary is typically indistinct, and the whole image fades rather rapidly, often disappearing toward its center, as adaptation to the new stimulus takes over. This can be accelerated by rapid eye movements or blinking and retarded by holding the eye stationary. There is thus little doubt that this sort of local adaptation is a physiological change in the sensitivity characterisitics of this area of the retina itself. There has been a change both in the total sensitivity, shown by the darkening, and in the chromatic sensitivities shown by the new color seen against the otherwise achromatic new stimulus. The complementary hue of this new response shows that the adaptation has been selective for hue, and the rough persistence of the degree of saturation indicates that adaptation has also responded to the purity of the original stimulus. It is found that the appearance of this afterimage is not dependent on the actual initial energy distribution involved; all metamers of the stimulus produce the same result. This adaptation process thus appears to follow, at least roughly, the same rules as those followed by the receptors in giving the original color perception.

In predicting the approximate color of afterimages it is convenient to think of both the primary and secondary stimuli in terms of their red-green-blue mixture metamers. Suppose the initial isolated stimulus is a 4° circular patch seen as a fairly saturated green. Such a stimulus contains little blue or red light and, accordingly, has little effect on eye sensitivity in these regions, leaving the eye in much the same state in these regions as does the no-stimulus surround. When the new and larger second stimulus with equal stimuli in all three regions is substituted at the same luminance, the region of the eye that has adjusted to green is of lower sensitivity for this color. The color seen is thus the achromatic stimulus with green removed, the roughly complementary "pink."

That this is the proper interpretation is easily shown by using a chromatic second stimulus. Suppose the second stimulus is cyan, that is, a mixture of blue and green. The afterimage will now be seen as blue. If the second stimulus is yellow (red plus green), the afterimage will be red. The perceived afterimage color is thus roughly predictable from this simple model. It must be kept in mind, however, that this description is in terms of a convenient metamer-type of stimulus; the same result is obtained for stimuli with the same chromaticities, regardless of their energy distributions, including, within limits, monochromatic stimuli. Thus adaptation to monochromatic green will make monochromatic cyan or yellow look bluish or reddish, respectively. The change is thus not a change in the effective energy distributions of the stimuli but in what we can call, for want of a better term, "effective receptor responses." In terms of hypothetical receptors, however, the change has to be thought of as a change in receptor output and not as a change in sensitivity distribution. This follows from the fact that any metamer can be substituted for the second stimulus without changing the color of the afterimage.

Since these facts imply that afterimages follow the regular color mixture laws and the effect producing afterimages may be thought of as local adaptation, there has been an assumption, usually attributed to von Kries, that adaptation in general can be thought of in these terms. This has been found not to be the case, at least in terms of any sensitivity distributions yet proposed. I shall consider the matter further in a later chapter, but we should note here that the rather obvious approach to discovering such a unique set of sensitivity distributions by a study of the actual color of afterimages appears never to have been pursued colorimetrically. It could be done, of course, only by inter-eye comparisons.

A distinctly different set of phenomena is associated with the so-called "Maxwell spot," and it is enlightening to consider it in the present connection; it also gives rise to "afterimages." Suppose that we have two large adjacent stimuli, both seen as achromatic, but one having a uniform energy distribution throughout the spectrum and the other consisting mainly of complementary bands of red and cyan light; and suppose the uniform distribution is on the left. (It is helpful if they are projected areas in a dark room.) If we look first at the left side we may or may not be aware of a poorly defined yellowish patch surrounding our point of fixation and moving as we move our eyes. This is the Maxwell spot, caused by the yellow pigmentation of the macular region surrounding the fovea. If we look *first* at the right side, a similar spot will be seen at once as distinctly reddish. On continued viewing of either side the spot rather promptly vanishes. The colors of the spots are due to the change in *energy distribution* of the light reaching this part of the eye, and the disappearance is due to local adaptation of the area of the retina beneath the yellow pigmentation has, in effect, destroyed the match between the two sides for this region of the eye. This can be demonstrated easily by holding a yellow filter over the eye. This makes the right side distinctly red compared to the left.

If we continue to look at the right-hand stimulus until the reddish spot disappears and then look at the left side, we now see a greenish spot (probably cyanish) because the local adaptation under the local yellow pigmentation has decreased the red sensitivity more than the blue and green. This spot will also disappear fairly promptly on continued viewing. It is instructive as to the general nature of adaptation equilibria to look back and forth at the two sides until a persistent spot is seen. This will appear reddish on the right and greenish on the left and, if the nature of the stimuli and their distance permits, can be seen as half greenish and half reddish by fixating on the dividing line.

The same two stimuli can be used to illustrate a point we mentioned earlier. If, for example, a fairly large, green stimulus is fixated, and Maxwell spot phenomena are overlooked, transfer of the afterimage from this stimulus to the dividing line between the achromatic ones will show the *same* pink color on both sides. The phenomena involved in these demonstrations thus nicely illustrate several quite fundamental facts. The last one shows that metameric matches, as matches, are independent of eye adaptation. The changing colors of the Maxwell spot show that in general metamers are destroyed by a change in their energy distributions. The stabilization of the spot with continued alternation of the stimuli is characteristic of all adaption; with time a *dynamic equilibrium* is always soon reached.

The disappearance of the Maxwell spot on continued viewing is a distinctly different phenomenon. It suggests that any *rigidly fixated* stimulus area (immobilized image) in a larger uniform area will disappear due to adaption and be *seen as the same* as the surrounding area. This is true but outside the realm of the present discussion.

SECTION TWO

RELATED STIMULI

In the preceding exposition it has been possible to rely almost entirely on the earlier literature for the facts involved. In the next few chapters many of the facts, although published, have a somewhat different status in that they are results obtained by myself and my colleagues and have not yet had the benefit of critical discussion by other workers. Furthermore, many of the studies on which conclusions are based were exploratory, intended to uncover the nature of the relations involved rather than to establish them quantitatively.

The studies themselves were the outgrowth of the discovery, published in 1959, of a previously unreported visual threshold, related to the perception of grayness in colors, which was encountered during investigations into the nature of saturation as a perceptual variable. The results reported were obtained intermittently over the years that followed, and the real significance of the threshold has become apparent to us only recently. In the following discussion I have attempted to give the facts in terms of our present concepts rather than in terms of our ideas as they evolved. I thus make use of data and investigations of widely varying precision and date and often rely on studies that, in retrospect, could have been designed far more to the point.

I do not think these facts weaken the conclusions that are reached. They do weaken the presentation because of the obvious gaps in the data where relatively simple experiments would support our conclusions. An attempt is made in the text to indicate the nature of such needed experiments and to indicate the relative reliability of the data from the various studies. Statistical analyses were not undertaken on any data, and all results, except the earliest ones, are based on observations by one or the other of only two observers. Our belief from the very start was that if we could once obtain data for the grayness threshold as it relates to a sufficient number of variables, its relationship to the established facts of colorimetry would become clear. It could then be established in terms of internationally accepted data. This has turned out not to be the case; thus the facts have to be presented in terms of our data.

The development of the problem, while still tentative as to many details, has involved acceptance of the fact that five independent perceptual variables are necessary to represent the general case of all related colors, a conclusion I have reached with reluctance as great as will be that of any of my readers. I am, however, convinced that it is true, that it clarifies many of the apparent anomalies in the subject, and that, if pursued beyond the point to which I have been able to carry it, it will establish relatively simple correlations between colorimetric and perceptual facts.

For the convenience of those interested in the development of the subject as published, I give below (as well as in the bibliography) the pertinent references. They cover most of the experiments described in this section, although some facts are presented here for the first time. All were published in the *Journal of the Optical Society of America.*

"On Some Aspects of White, Gray and Black," **39,** 774, September 1949.

"Fluorescence and Gray Content of Surface Colors," **49,** 1049, November 1959.

"Variables of Perceived Color," **54,** 1467, December 1964.

"Luminance and Induced Colors from Adaptation to 100-Millilambert Monochromatic Light," Letter to the Editor, **57,** 279, February 1967.

(With B. Swenholt) "Chromatic Strength of Colors," I. "Dominant Wavelength and Purity," **57,** 1319, November 1967; II. "The Munsell System," **58,** 580, April 1968; III. "Chromatic Surrounds and Discussion," **59,** 628, May 1969.

Six
ACHROMATIC RELATIONS

In the first section we considered the nature of the light stimulus that produces the perception of color; it can be described completely by a statement of the energy it contains at each wavelength to which the eye is sensitive. There are thus an infinite number of possible color stimuli. Because of the finite discrimination of the eye for changes in this energy distribution, this does not lead to an infinity of different colors. The number, however, would be extremely large if it were not for the fact that a great many of these distributions look alike to the eye.

This is the subject of metamerism, large groups of stimuli can be found in which all stimuli appear identical. Individual observers are found to be quite different, at the extremes, in the stimuli they assign to a particular group; but a high percentage of people are approximately alike in this respect, giving rise to the concept of a "normal" observer, comprising within its "normal" variability some 95% of the population. In order to get a grouping of stimuli useful for colorimetry it has been necessary to agree on a Standard Observer with average characteristics. With this as a basis it is found possible to group all possible stimuli in such a way that each group of visually identical stimuli can be identified by means of only three suitably chosen variables. Ths simplification, along with the finite discrimination of visually different stimulus groups, reduces the infinity of physically different stimuli to a manageable number (in the millions), each describable by amounts of three (psychophysical) variables. These three variables thus describe *all the physically possible* single light stimuli. They can be thought of as enclosed in a space limited in two of its dimensions by the physics of the stimulus and in the third by the sensitivity and tolerance limits of the eye.

Since this space includes points that represent all the possible stimuli, all possible color perceptions must arise from the effects of one (*or more*) such stimulus. In other words any complex spatial array of stimuli is always describable, point by point, by the instantaneous values of these three variables, and the resulting perceptions must be ultimately traceable to their combined effects. It does not follow, and is not true, that there is a unique perception simply related to *any* given point in this stimulus space, including those on the boundaries.

Understanding color perception thus involves starting with the simplest possible stimulus and gradually increasing the number of stimuli seen simultaneously (or successively), noting at each stage the number and nature of the color perceptions produced and their dependence on these three variables in each stimulus.

Up to this point we have considered single stimuli seen as uniformly distributed over a moderate visual angle and isolated, that is, in the complete absence of any other light stimuli. We have found the three main perceptual variables that we called hue, saturation, and brightness, but we noted an apparent anomaly in the fact that *isolated* stimuli of the same luminance but high purity could not be described as being equally bright on an absolute (introspective) basis. We also noted that immediately successive stimuli showed that the response of the eye to the second stimulus had been modified by the first.

We proceed in this section to consider the perceptions produced when two (and finally three) stimuli are present at the same time and start, in this chapter, with the simplest case of two, in which both are seen as achromatic. We shall encounter semantic problems that will require a lengthy digression that will considerably clarify both the nature of the perceptual variables of color and my definition of the word color itself.

Suppose that, for descriptive purposes, we consider an isolated 10° circular stimulus, isolated from any external stimuli, at a luminance of 100 ml and seen as achromatic but having at its center a smaller circular stimulus (1 to 2°) having the same chromaticity but variable independently in luminance only. It is apparent that, at matched luminance, these stimuli will be metameric; and, ideally, the central stimulus will disappear as such, that is, will be indistinguishable from the surround. The whole under these conditions will be a single isolated stimulus producing zero saturation (hence what I shall call zero hue) and characterizable by a single absolute (introspective) brightness. This brightness would increase or decrease if the luminance of each was increased or decreased by the same amount.

If, however, we decrease the luminance of the central stimulus only to, say, half its value, we at once encounter a new perceptual variable not before seen from any stimulus so far considered. This new perception is called "grayness' in ordinary speech and is, perceptually, on a par with hue, that is, it is a positive perception, describable only by reference to common experience. But this is not the only perceptual change that has been introduced; we can now also describe the stimulus as "darker" (less bright) than the surround. We shall find later that when hue is present these are independent variables. For the present case, they are not, and we can consider them together. Note that we have now extended the meaning of the word "achromatic" to include all color perceptions in which no hue is seen. It is in this sense that the word "colorless" is used in ordinary speech, but we will include in the term "achromatic" the perception of grayness and use "colorless" to mean the absence of both gray and hue.

As we continue to decrease the central stimulus luminance, increasing grayness is seen, and the darkness of the stimulus increases simultaneously with respect to the surround. Eventually a point is reached where common usage would call the perception "black." This is not a definite point; there is no threshold which can be generally agreed on, below which it is clearly black and above which it is clearly gray. There is, rather, a range over which black is a better description than dark gray, and there are distinctly varying degrees of blackness which can be seen by varying the stimulus. A luminance point is eventually reached where a further decrease in luminance produces no further perceptual change, either in blackness or darkness. This point can be set with fair precision as the point above which a change can be seen and below which it cannot. This is obviously the luminance threshold for light perception under the circumstances. I shall occasionally refer to it as the "black point,"meaning the point where black has reached a *maximum* for the conditions.

This whole series is somewhat analogous, as far as the gray-black variable is concerned, to the series discussed earlier in which the purity of a stimulus was changed from 0 to 1. In that series hue appeared and increased to a maximum limited by the physics of the stimulus. Here gray appears and increases to a maximum that is limited by the characteristics of the eye. Here, also, darkness increases to a maximum at the black point, but this series is *not* analogous to the brightness series produced by decreasing the luminance of an isolated stimulus. For the isolated stimulus, a decrease in luminance simply made the stimulus appear "dimmer" without the introduction of a new perceptual variable. The words "darker" and "dimmer" are not synonymous. "Darker," as used here, means "darker than the surround" and so is a relative term describing the relationship of the two stimuli as seen, whereas "dimmer" is a judgment based on memory; they are different perceptions. They can, in fact, both be seen in the case under consideration. If *both* stimuli are decreased proportionally the whole field becomes dimmer, but the darkness of the central stimulus, as we are using the word, does not change appreciably at the level we are considering. Unfortunately, in ordinary speech, "darker" is often used in place of "dimmer," as when we say that one part of a room is darker than another, and the phrase "in the dark" usually means in very dim light. The achromatic series from colorless through degrees of grayness to black is also, almost universally, referred to in the literature as the "white to black" series. There are thus serious semantic difficulties, due mostly to poorly considered analogies, and it is necessary to consider in closer detail all the words and concepts involved.

WHITE

We can best start with "white," especially since I want to place a restriction on its meaning. In ordinary speech "white" has nearly as many meanings as the word "color" itself, but even if we restrict ourselves to its meaning with respect to perceived color it still involves at least three concepts, to any or all of which it may refer simultaneously. These are: (1) the absence of hue, (2) the absence of grayness, and (3) the presence of diffusion that scatters the incident light in all directions or, more generally, the fact that light is seen to reach the eye uniformly from all parts of the stimulus. The first two have to do with the perceived color; the third is the perception of a stimulus characteristic *not* related to color.

So far as I know "white" *always* means either that the observer is not conscious of the presence of hue or grayness (although the word is often used where critical observation would reveal either or both) or that he does not attribute the obvious presence of hue to the stimulus of which he is speaking. Although we will discuss the matter at length in a later section, it is instructive to consider here the use of "white" with respect to objects. For objects like paper, tile, paint, clouds, and smoke, it carries the added meaning of "effectively opaque." However, the word is also often used to refer to clear, colorless glass. To me this is a poor use of language because it is used simply to avoid the longer expression and then requires the use of an even longer expression if a nontransparent, diffusing but colorless glass such as "opal" is to be described. Such usage does, however, call attention to the fact that there are various degrees of whiteness for which we have no common speech terminology.

For example, series of whitenesses can be generated by suspension of not too finely divided chalk particles in water. Such a series (assuming it to be without hue) runs all the way from opaque white through various degrees of whiteness to zero whiteness at pure water. In no member of the series is grayness apparent. The perception involved is a change in the extent to which the object diffuses the light. The same series may be generated by grinding the surface of a mirror, by forming concentrations of colorless smoke, and so on. (See Evans 1949.) We have, for such series, descriptive terms other than "whiteness," such as "turbidity," "clarity," and "degree of diffusion," none of which have the added connotation of colorless.

Since two whites can be compared for whiteness it is apparent that if we are to consider "white" as a color term, whiteness must be taken as an *independent* variable and that there will then be separate color domains, each having a different degree of whiteness. This would add another variable to perceived color. For this reason it seems preferable to me to say that whiteness is *not* a color variable but that it refers to a combination of stimulus physical characteristics that have in common the fact that they are *colorless.* On the other hand, this does not mean that "white" is an inappropriate term to use in reference to the stimulus when it does have the combined properties of appearing achromatic *and* diffuse. It seems to me that the use of the expression "white light," which is almost universal, is on the same level as "white glass"; both are bad terminology. It is perfectly appropriate, however, to describe our 10° achromatic surround as white because it *does* have these characteristics. To summarize, then, I consider the zero point of color perception to be "colorless" and not "white"; it is exemplified by pure air over short distances or completely transparent glass in which no reflections can be seen; that is, zero color means *no* color is visible.

We might note here that this approach to the perception of white places it in the same class as many other appearance attributes such as glossiness or transparency, any of which can be present in a perception stimultaneously with any color but which are not, *per se,* part of the color perception itself. This is discussed in more detail below; it restricts the meaning of "perceived color" to those perceptual attributes that can be changed by changing *only* the energy or the spectral distribution of the light reaching the eye from some part or all of the complex stimulus being viewed. It recognizes color as a closed subdivision of the larger subject of appearance.

GRAYNESS

Returning now to the perception of grayness and considering how it was produced in our example, we find that it is not only a new perception but that it has a new kind of relationship to the stimulus. In the case of an isolated stimulus we could relate the perceptions to the physics and psychophysics of the stimulus alone. Here we have a perception that appears to be due to the stimulus but is in fact independent of it. This independence is evident from the fact that exactly the same series from a metameric match through all the grays, down to and including the black point, can be generated either by reducing the luminance of the primary stimulus only or by *increasing* the luminance of the surround only. Grayness is a perception that is due in fact to its relation to other stimuli and not solely to its own characteristics. So true is this that in the paper cited earlier I wrote that "grayness is the perception of relative luminance." We shall find that this statement applies only to what we have so far called "darkness" and that grayness, while still dependent on luminance relations, has a different relation to the stimulus when it is chromatic. It is, in fact, the reason *the* two perceptions are independent in the general case.

Hering (see Hering 1905–1920) introduced the concept of a "veiling" of the stimulus by the surround excitation, and this is a good description of what occurs. (He also speaks of a veiling by white with which I obviously do not agree.) Presumably what occurs is that the sensitivity level of the eye is maintained by the larger and brighter surround, and the eye is thus not able to increase its sensitivity to accommodate the decreasing luminance of the stimulus. Increasing grayness can thus be thought of as a decreasing *effectiveness* of the stimulus to produce the perception of light. The very positive character of black as a perception—as positive as the high-purity hues—remains to be accounted for. If we recall, however, that the eye spontaneously increases in sensitivity to adjust to low overall luminances and that in the absence of any light stimulation sensitivity reaches such a high level that "noise" (the so-called "ideoretinal light") is produced, it is not surprising that *total* lack of response is seen as a very positive fact. In any case, I shall take the position that grayness (and darkness in this case) is the perception of the relative effectiveness of the stimulus to produce the perception we call light, that it grades continuously through the series of grays, and that black is the perception of total inadequacy. It has nothing to do with the chromatic or other perceptions that may accompany it.

We have noted that white is to be considered an appearance characteristic not included in the concept "color" but that gray, black and dark are included. It is worthwhile to consider somewhat more completely this relationship of color to appearance.

STIMULUS APPEARANCE CHARACTERISTICS

In 1911 there appeared in Germany a book by David Katz, called *Der Aufbau der Farbwelt* and the second edition (1930) was published (1935) in English as *The World of Colour.* This book presents a masterly introspective analysis of the possible perceptions involved in stimuli for color and has had a great influence in clarifying the relationships between the physical stimuli and the perceptions produced. In particular, it had a deciding influence on the resolution of the twenty-some-year deliberations of the Committee on Colorimetry of the Optical Society of America. Their final report, published as *The Science of Color* (1953), in essence accepted Katz's categorizations. I shall take their published form of it as the basis for the following discussion. Note that I do not actually disagree with any part of it but that I regard a number of its implications (and many of the deductions that have since been drawn from it) as distinctly unfortunate for the subject of color perception.

Basically the approach divides all possible color perceptions into 3 "Attributes of Color Sensation."—brightness, hue, and saturation—then specifies 5 "Modes of Appearance," and lists 11 "Attributes of Modes of Appearance," not all of which apply to each "mode." The list of the modes and attributes is given below, the numbers following each mode referring to the applicable attributes (1–5 means 1 through 5 inclusive, etc.).

Modes	Attributes
	1. Brightness (or lightness)
Aperture (1–5)	2. Hue
Illuminant (1–8)	3. Saturation
Illumination (1–3)	4. Size
Object modes:	5. Shape
Surface (1–11)	6. Location
Volume (1–9)	7. Flicker
	8. Sparkle
	9. Transparency
	10. Glossiness
	11. Luster

Nothing would be gained, as far as the present work is concerned, by a detailed analysis of these lists. They serve admirably, however, in clarifying my position with respect to the word "color," especially as modified by the adjective "perceived."

A number of facts are evident immediately from a brief study of these lists:

1. All of the attributes following the first three are perceptions of space-time characteristics of the stimulus having nothing to do with its spectral energy distribution.
2. Grayness is not recognized as a perceptual attribute except as it is, perhaps, suggested by the parenthetical "or lightness."
3. The five listed modes are limiting cases of five types of situations *and* stimuli without a distinction between the two.

The first three attributes, of course, reflect the generally accepted belief that the perceptions (called sensations) arising from the stimulus must be due to the variables of the stimulus *itself* and so must be three in number. The "or lightness" is required to maintain three variables in the Object Modes. It is our thesis that this supposition is inadequate.

It may be helpful at this point to consider item 3 in terms of the organization of the present book.

Although "Modes of Appearance" is an adequate translation of "Erscheinungsweise," I find the term misleading; it carries the connotation that the appearance of the stimulus itself changes as a whole in the different situations. We shall find it a better description to say that the perception of the stimulus sometimes *splits* into two or more *separate* perceptions depending on the perceptual framework to which the observer relates a specific aspect of the stimulus. In other words, it is not correct to assume that the same stimulus cannot produce more than one color perception at the same time. With this in mind we can consider the modes separately, noting that some will be discussed in far more detail later.

The Illuminant Mode includes specifically all cases in which the stimulus is seen to be a light source and yet is much broader in what it includes than this implies. It includes the case of isolated colors with which we started, and which we would not ordinarily think of as illuminants since they do not illuminate anything. It also includes any stimulus in a complex field that is very much brighter than the other stimuli but too small to have much effect on overall eye sensitivity. On the other hand it does *not* include stimuli *known* to be light sources, such as some indicator lights, when their luminance is at or below the average luminance level of the scene. It is thus difficult to find an expression that really describes this class of perceptions. We will discuss it again when we consider adaptation. In effect the Illuminant Mode includes the perception of all stimuli that produce the perceptions characteristic of isolated colors and so perhaps should be thought of as the "isolated stimulus mode."

Perhaps the greatest contribution Katz made in his book was his insistence that the light that is seen to be illuminating objects is perceived as *separate from* the objects. This is a concept that is obvious to the naive and obscure to scientists; it is a basic tenet of the present book that this is so fundamentally true that no perception of a complex scene can be analyzed correctly without taking it into account. It is elaborated in a later section and is the situation in which the "splitting" of perceptions, mentioned above, most frequently occurs.

The Illuminant and Illumination Modes thus nicely separate out two major classifications of color perceptions. We shall find that in each of them the number of possible color perception variables is different. We have already found that there are four for isolated colors and will find five for illumination. It is the separation of the remaining modes into three groups and particularly the implication that "aperture colors" are somehow different in their *color* variables than the others that seems to have done so much harm to the subject of color perception. In fact they are all the same case as far as color *per se* is concerned, and all require five independent variables for their discription. The logic behind their separation is apparent enough; what is not apparent is that it is not based on color. Thus the reason for separating object colors into surface and volume (volume meaning transparent) is that white is included as a color variable in surface colors. If white is considered an object appearance rather than a color perception then the two cases are the same. Again, the reason for separating out aperture colors has nothing to do with color; the reason is that an aperture stimulus *is not limited* by the physics of objects. In the cases of surface and volume colors the stimulus is restricted to a reflectance or transmittance at any wavelength of 100%, and the attainable purity is limited by this fact as well as by the relative luminance. In the aperture case these limitations do not apply; relative luminance may have any value, and purity can be unity. Over the range in which the stimuli *can* be the same the *color* perceptions are identical. All three cases are included in the concept of related colors; that is, stimuli seen in relation to each other, and will be so treated here. They all can be considered to have five independent, perceptual variables. From this standpoint, the Aperture Mode is the general case that *includes* the Object Modes. We cannot, however, discard the concept of object perception as such because it is the basis for the simultaneous perception of illumination and objects as separate. In the more general Aperture Mode this does not occur. Thus the distinction is between situations and not colors. Basically, the problem of the classification of color perceptions is purely semantic, calling for considerable redundancy. It is not possible to classify all color perceptions as isolated or related, as was attempted by Ostwald, because illumination perception is neither. Nor can object colors be separated out, because they are included in aperture colors. The real distinction comes in a combination of the frame of reference in which the observer is perceiving and the possibilities for perceptions offered by the stimuli; neither can be neglected.

In the present section, I shall restrict the discussion to homogeneous stimuli, seen in relation to one or more others, hence in the Aperture Mode. Where it is necessary to consider illuminated objects I shall consider the illumination completely uniform, achromatic, and of known energy distribution. It thus enters as a determinant of the stimuli only and can be neglected perceptually except for absolute intensity. Illumination perception as such will be discussed in Section III.

LIGHTNESS

In our example of the two achromatic stimuli with decreased luminance in the central area we noted that the central patch could be described as darker, as well as grayer. We also noted that the term "darker" is ambiguous. This is the "or lightness" attribute and the problem thus glossed over is fundamental to our subject. If the belief is to be maintained that there are only three variables of perceived color, then somehow the brightness attribute of isolated stimuli must be made to correspond to some attribute in the related case. This must be done because hue and saturation are obviously the same variables in both cases. So far as I know, it has been assumed throughout the entire literature that the correlate of brightness is the perception that we have just described as a combination of grayness and darkness, and this in spite of the fact that, introspectively, they are both very different from brightness. The term "lightness" was introduced, basically, to acknowledge the fact that the perception is different, but the anomaly involved was not officially recognized. This treatment is in complete accord with all of the literature with which I am acquainted; it was not introduced by Katz. The statement usually made is that "in the case of object colors the variable corresponding to brightness is lightness" or the equivalent in other terminology. "Lightness" is thus made to carry the burden of maintaining only three variables by adding on to brightness the perceptions of grayness and what I have so far called "darkness." In fact these are not only three different perceptual variables but are also independent of each other in the general case. The requirement is, of course, to abandon the belief that there *are* only three variables. Fortunately we do not have to change the terminology except to separate out the concepts of brightness and grayness from that of lightness, thus leaving only what I have called "darkness." The obvious pun that this makes lightness the same thing as darkness illustrates my objection to both words. They both represent artificial restrictions on two words from everyday speech. However, I do not know of better words for the concept and in any case hesitate to coin new words unless necessary.

Specifically, then, the three perceptions involved will be treated in the following manner. "Brightness" I will use for the perception of the overall effective intensity of the stimulus or stimuli. In the immediate case it refers to the perception of the general luminance level of both the surround and the central area. "Lightness" I will use for the perception of the apparent luminance of the central spot *with re-*

spect to the luminance of the surround. It is thus the perception of the *relative* apparent luminance of the central stimulus with respect to the surround. Since in the case of reflecting surfaces this relative luminance is called "reflectance" and for transparent objects "transmittance," we can also say that lightness is the perception of apparent reflectance or transmittance. I shall have to postpone discussion of the term "grayness" to the next chapter because there I shall want to include it as one aspect of a more general perception that I will call "brilliance." I thus take the position that perceived color has five separate variables: hue, saturation, brightness, lightness, and brilliance.

THE NONCOLOR ATTRIBUTES

Before entering into details elaborating this position, it may be helpful to consider briefly the appearance factors that are not being included in the concept of perceived color. They are often more important to the observer than the color. They also often confuse the comparison of colors both by competing for the observer's attention and by actually modifying the color stimulus without the observer being aware of their doing so.

In the order listed in the OSA table, *size* and *shape* are obvious geometric characteristics of the stimulus; they change the distribution of the areas of the retinal image and hence modify the effectiveness both for direct perception and for the effect on eye sensitivity. Thus we find that a long narrow line appears different from a small circular area and a small circular area different from a large one. *Location* has been added as an attribute because it is necessary when a distinction is made between aperture and surface colors. Actually it means that the stimulus is nonuniform when it can be located visually, because it is only when there is a micro- (or macro-) structure present that there is anything on which to focus. This is an important point. In the elementary subject of color perception it is necessary to assume that the stimulus being discussed is completely uniform over its area because we do not know how the eye integrates nonuniformities. But a *perfectly* uniform area as part of a real surface would be nonlocatable except perhaps by touch and would become an aperture color. It illustrates the artificiality of the distinction and also the fact that the *color* perceptions are the same.

Flicker and *sparkle* are obviously time variations of the stimulus that may or may not vary anything but its luminance. Even if luminance is the only change, however, the time response of the eye can lead to changes in the perceived color so that, again in an elementary treatment, it is necessary to assume the stimulus to be static. There is no indication, on the other hand, that either transitory or cyclic stimuli introduce any *new* variables into color perception. *Transparency* we have considered in connection with the concept of white. There it was required only because of the unnecessary separation of object colors into surface and volume.

Glossiness and *luster* are somewhat different from the other attributes in that the observer can be misled by them without being aware of it. For our purposes we can describe glossiness as due to high nonselective outer surface reflection and luster as due to a stimulus that has different spectral energy distributions or luminances when seen from different directions (hence different for the two eyes). Both cases deal with more than one color *attributed by the observer* to a single stimulus. The color seen will depend, therefore, almost entirely on the extent to which he either discounts one or the other mentally, or tries to combine them into a single perception. An instrument will necessarily read some fixed relation between the two (possibly more than two for luster), but an observer will tend to reach some sort of compromise opinion.

Thus it appears that none of the attributes below the first three on the list introduce any new perceptual color variables but instead rather completely summarize possible complexities that may arise in the nature of stimuli that are *not* uniform in space or time or both. On the other hand, the observer will find it extremely valuable to be on the lookout for them if he is comparing two stimuli of different nature. No match for color can have real meaning unless the nature and amount of these modifying effects are *specified.* Thus a large area of painted wall may appear different from a small patch of the same paint, for reasons quite aside from interreflections; and a glossy color may appear more or less saturated than a matte one, depending on what it is reflecting and how much the reflection is discounted. For really acurate visual color matching *all* attributes listed should be the same for both samples. From the point of view of color perception all the attributes except the first three and transparency are eliminated by the assumption of stimuli that are uniform and static with respect to space and time. Transparency involves the question of white, which was discussed earlier. With white excluded as a color perception we can say that within limits any color perception may be accompanied by the perception of any of the attributes.

SUMMARY

The introduction of a second stimulus into the viewing field not only presents new perceptions but also raises a number of semantic problems. Discussion of these has led to a rather discursive treatment that tends to obscure the main points. Since some of the positions taken are necessarily arbitrary and will be adhered to in subsequent chapters, the following brief notes on the various terms may be helpful to the reader. They are not intended as rigid definitions. In order to make the list more useful for reference I have added several terms from later chapters.

Achromatic. When a light stimulus produces a perception that is without hue, the perception is called achromatic. It thus includes the perceptions of gray and black.

Appearance characteristics. The perception of objects or, in the limiting case, of just two uniform stimuli, may involve many perceptions not peculiar to color but necessary perceptually in the process of distinguishing parts of the scene. All of these are included in this term, along with color.

Black, blackness. Black is considered the limiting case of grayness. It reaches a maximum, under a given set of conditions, at the point where the perception becomes totally independent of both the energy distribution characteristics and the intensity of the stimulus. A whole range of very dark grays is usually called black; two stimuli in the same scene can thus differ in blackness. The maximum attainable blackness depends on the scene, not the stimulus.

Brightness. This is taken to be the perception of the general absolute luminance level of the combined stimuli visible to the observer. It correlates with this luminance level always, in the sense that increasing it would make the scene look brighter and decreasing it would make it appear less bright. Memory comparisons between different scenes at different times may not correlate well, but brightness always refers to the apparent luminance level.

Brilliance. This is a term I introduce later to cover the perceptions of both grayness, as discussed here, and fluorence, as discussed in the next chapter. Its independence of both lightness and brightness is discussed in Chapter 8.

Chromatic. This is used to characterize all color perceptions in which hue is present.

Color (perceived). The word color by itself is intended to refer only to the perceptions caused by a uniform, static, light stimulus. It is modified by "perceived" wherever there seems a possibility of misunderstanding. Thus the usage is not intended to exclude other definitions but rather to indicate the closed subdivision of the general subject of appearance encompassed by the five variables, hue, saturation, brightness, brilliance, and lightness.

Colorless. From the usage of the word color it follows that this term implies the absence of, or zero values for, the five attributes of color. Since a brightness of zero would mean no perceptions at all from a stimulus and this could be construed to include black, which is not intended, its usefulness is restricted to indicating the absence of all attributes except brightness. In this sense the concept is included in "white."

Darker. This comparative term is intended in the common speech sense only and means either "less bright than" or "less light than," depending on the context.

Dim, dimness. These are used as the opposites of only bright and brightness.

Fluorent, fluorence. These coined words are applied to the perception that has come to be associated with stimuli that are physically fluorescent. It occurs when the brilliance of the stimulus exceeds that of its surrounding or comparison stimuli (Chapter 7). The words are justified by the fact that fluorescence of the stimulus is neither a necessary nor a sufficient requirement for the perception.

Gray, grayness. These terms apply to the unique perception generated when one stimulus is seen to be less brilliant than its surround or comparison stimulus or stimuli.

Lightness, lighter. Both of these are comparative terms. Lightness refers to the apparent relative reflectance or transmittance of a stimulus considered as a reflecting or transmitting object regardless of its physical nature. It thus applies to aperture as well as to object colors and is directly related to relative luminance.

White, whiteness. These terms are reserved for those cases in which the stimulus sends light uniformly to the eye from all points of its area *and also* produces a colorless perception in the above sense. In objects this requires a uniformly diffusing material. White, therefore, is taken as a combination term that includes both perceptions and not as a term unique to color. The word "whiteness" is made necessary by the existence of varying degrees of diffusion, down to zero, which do not necessarily alter the colorless aspect of the perception.

Whiter. The comparative form thus refers to the diffusion aspect and not to the colorless aspect. In common speech, however, it is often used to make a distinction between two stimuli both seen as white but with different brightnesses. I have used "brighter" and "dimmer" for these cases to avoid confusion.

Zero gray. Grayness and fluorence are mutually exclusive perceptions from a single stimulus and neither is necessarily present. I introduce this term in the next chapter to indicate a perception in which both are absent.

Seven *CHROMATIC STIMULI WITH ACHROMATIC SURROUNDS*

In the previous chapter we considered the case of a stimulus seen as achromatic with a larger surround of the same chromaticity. We found that lowering the luminance of the stimulus relative to the surround introduced two new color perceptions, grayness and lightness, that varied together from the match point to black. In the present chapter we consider chromatic stimuli but again surrounded by a stimulus large enough and bright enough so that it remains achromatic. In examples taken from our own work the surround is at 100 mL and subtends about 10° and the stimulus about 1° with its luminance independently variable. We want to consider the perceptions produced in this case as relative luminance is varied, but this time, for convenience, I start with black.

If the central stimulus is at a very low luminance with respect to the surround (say $<1/1000$), black is seen regardless of the chromaticity of the stimulus. Suppose that the stimulus is monochromatic light with a wavelength of 700 nm. As the luminance of this stimulus is increased black persists without change until a point is reached where hue becomes just perceptible. This is the "black point" and corresponds to that of the previous case. The perception just above this point is that of an exceedingly faint hue (in this case reddish) seen in the blackness. As luminance is further increased hue becomes stronger and blackness simultaneously decreases until, as before, it becomes more descriptive to call this component dark gray. Above this region both changes persist, the grayness decreasing and hue perception (hueness?) becoming stronger until a fairly well defined point is reached at which grayness has disappeared. Lightness, of course, has increased continuously from zero at black. At this point, the perceived variables are hue and lightness and overall brightness. For λ 700 this point occurs when the stimulus luminance is of the order of $1/10$ that of the surround. This point can be found with good precision for all stimuli seen as chromatic; and I shall call it the "zero gray" point and designate it by G_0.

As we continue to raise the luminance above this point, lightness and the strength of the hue perception continue to increase, and a new perception appears and becomes stronger. This new perception can best be described by saying that the central stimulus now appears as though it were fluorescent, and, again, this occurs for all stimuli seen as chromatic, and always occurs while the stimulus luminance is still *below* that of the surround. G_0 therefore, is a *threshold* between stimuli that appear grayish and those that appear to be fluorescent as the luminance is changed relative to the surround. This threshold was apparently new to the literature when it was found by my colleagues and myself (Evans 1959). Since no actual fluorescence is involved in the phenomenon and it seemed desirable to call attention to this fact, I coined the term "fluorence" for the phenomenon, saying also that the stimulus is "fluorent" under these conditions. G_0 can thus be described as the threshold between stimuli that are grayish and those that are fluorent, and I shall use it in this sense throughout. For the reason that it can be demonstrated that these two perceptions are mutually exclusive and for additional reasons to be given presently, I shall consider that grayness and fluorence are different aspects of a single perceptual variable I shall call "brilliance;" it may be considered negative for grayness and positive for fluorence, or simply continuous from the black point.

With continued increase in relative luminance, lightness and fluorence increase until a point is reached where the lightness matches that of the surround. This point coincides more or less accurately with a luminance match between the two. Fluorence is quite strong at this point for high purity stimuli, and some experience and skill are required to recognize and set it with precision. This is the "heterochromatic brightness match point against white" that was discussed earlier and that the early investigators learned to set; they found it in accord with matches made by the "flicker method." Such matches were made by many skilled observers on our equipment and never failed to confirm our calculated luminance match points within individual differences. We also learned to recognize it ourselves but made no detailed investigation of it. It is apparent that above the G_0 point, where both lightness and fluorence are increasing, lightness as a separate perception becomes more and more difficult to recognize as such, becoming more and more obscured by the increasing brilliance. At low purities, where the fluorence is less intense, it is easily recognizable all the way. I shall consider, therefore, that lightness starts at zero at black and increases steadily with luminance until it matches that of the surround at a luminance match. For this reason I shall consider it tc be the perception of relative luminance. We will find in a later chapter that it can be varied independently of brilliance for the same stimulus chromaticity and hence must be considered a separate perception.

If we now continue to increase the luminance of the stimulus above the match point, two different perceptual phenomena occur, more or less simultaneously. For a range of luminances that depends both on the purity and the dominant wavelength of the stimulus, fluorence continues to increase and then rather abruptly decreases and disappears as a perception. Description of the perceptual change that takes place at this point is difficult, because it depends largely on the semantic background of the observer. One can say alternatively that the appearance of the stimulus changes from that of a surface color to that of a light source, that the stimulus itself becomes the reference point with the surround now appearing of lower lightness, or that the stimulus has now taken on the attributes of an unrelated color. I shall consider these all equivalent statements and suggest that what occurs is that the stimulus itself becomes at this point the main controlling factor in the sensitivity level of the eye (i.e., starts to control adaptation level). Above this point, and on up to the visual tolerance limit, the perception has the three main attributes given by an unrelated stimulus, hue, saturation, and brightness.

I have considered this relative luminance series in some detail because it is fundamental; it occurs for any pair of stimuli which are seen to be chromatically different. A G_0 threshold always exists and can be set visually. I have not considered hue because I cannot do so usefully at this point and because the whole subject of hue will have to be discussed separately. The rest of this chapter will be devoted to a discussion of the value of G_0 as it is affected by changes in the psychophysical variables of the stimulus itself, with the surround achromatic. All of the original data are, necessarily, from our own experimental work. It might be well to note briefly the philosophy and instrumentation involved in our results. From the very start our main interest was to discover the relationship of the G_0 threshold to the physical and psychophysical variables involved. Emphasis was placed on accuracy of the stimuli and on the variety of situations studied, rather than on precision of setting. Although all settings were made with great care and some series were run many times to reduce variability the emphasis was always on the nature of the function involved rather than on its quantitative determination.

INSTRUMENTATION AND APPROACH

Several instruments were used in the work, all of which are described in the articles cited at the beginning of this section. The final instrument was designed specifically for determining the dependence of G_0 on wavelength and purity, against both achromatic and chromatic surrounds, at a luminance which would be well up into the photopic region. Except as noted, all results were obtained with the surround at 100 mL whether chromatic or achromatic. To work at these levels easily, we used a filter rather that a spectrum-type instrument. Filters (mostly of the carefully blocked interference type) were obtained and carefully calibrated (to 3000 K). These filters were selected to transmit 31 dominant wavelengths in the spectral and extraspectral regions, more or less equally spaced by wavelength; all had excitation purities greater than 0.95 and were mostly at a purity of 1.0. Purity in the central stimulus could be varied from 0 to that of the filter, and any filter could be used (at full purity only) in the surround. Luminance of both beams could exceed 100 mL by a considerable amount. The main limitations, therefore, were inability to vary wavelength continuously and inability to vary the purity of the surround.

All settings were made monocularly by one or the other of two observers. The central spot was never fixated during adjustment, the intention being to keep the fovea adapted as much as possible to the surround, but the final decision was based on direct viewing. In all cases, one "reading" consisted of the simple average of four such "settings." The luminances of the stimuli were controlled by photographic silver circular wedges calibrated in density (log $1/T$) units, the actual luminances being calculated from the known beam luminances and the transmittances of the filters. The averages were thus average density settings. The precision of setting was characteristic for this type of visual threshold, quite good within one set of four and quite variable from day to day and between the different spectral regions. Where needed and when time was available, therefore, precision was increased by repetition in sessions that were days or sometimes weeks apart rather than by increase in number of settings in one session. The investigations were terminated by my retirement, leaving many unanswered questions, as will be noted. Also the sequence of facts as here presented is dictated by our present insight into their meaning and bears only a superficial resemblance to our thinking at the time the experiments were run. Some experiments that we would now consider crucial were therefore run out of curiosity, and there are many that we wish we had run and did not. It was only toward the end of the work that we began to realize we were dealing with a new, independent variable, and only recently that we have come to appreciate its essentially revolutionary nature, involving as it does a new (although dependent) psychophysical function.

The light source in our instrument was a coil filament projection lamp calibrated and maintained at a color temperature of 3000 K, giving an unfiltered intensity in the central stimulus of somewhat more than 10^6 mL. Light from this same lamp illuminated the surround and was filtered to a chromaticity of $x = 0.303$, $y = 0.326$ (correlated color temperature 7143 K). This was intended to approximate Ill. C, and while we regretted later that it was not closer, we thought it better to have all data comparable; it was also used as the achromatic component in the purity series. All the earlier observations were made by me (RME) and all the later ones by my associate throughout the studies with this instrument, Bonnie Swenholt (BS). A rapidly worsening cataract (which probably affected some readings) made it unwise for me to continue observing. Cross checks over the years showed our results to be very similar, and it is unlikely any conclusions were affected by the change of observer.

At first all of our attention was devoted to developing skill and experience at setting G_0 and determining its dependence on wavelength at the full purity of the filters. For most wavelengths the proper setting is fairly obvious. Since the stimulus is fluorent above and grayish below, a setting can always be made in the center of the relatively short range over which neither can be seen. In some regions of the spectrum, however, difficulties were encountered, which gave us much uncertainty. Yellow under the circumstances is never strongly fluorent and the tendency was to set at too high a luminance. For green it is difficult to decide between the darkening of the green and the introduction of gray; thus readings are quite erratic and tiring. In the far blue a quite different phenomenon takes place which again confuses the G_0 point: the appearance of violet in place of blue as the luminance is raised from below G_0. In other words, at short wavelengths there appears to be a second threshold quite close to the G_0 level in which a red component appears and produces the color violet from what before was blue. This problem is confounded by the fact that the actual luminance at G_0 is around 1 mL and so difficult to see clearly anyway. (This separation of a red component from violet occurred in many situations and needs further investigation. We finally had little doubt in our minds that violet consists of two independent responses, one blue and the other red.)

The final result of this work was the curve of Figure 7-1, showing G_0 in terms of the wavelengths and complementary wavelengths of our filters (corrected for filter purity). A word about the ordinate units in which G_0 is expressed is necessary. Because our attenuating wedges were calibrated in density (= log $1/T$ where T is the transmittance) it very much simplified our calculations to work in logarithmic units and, in fact, to leave them in terms of density units.

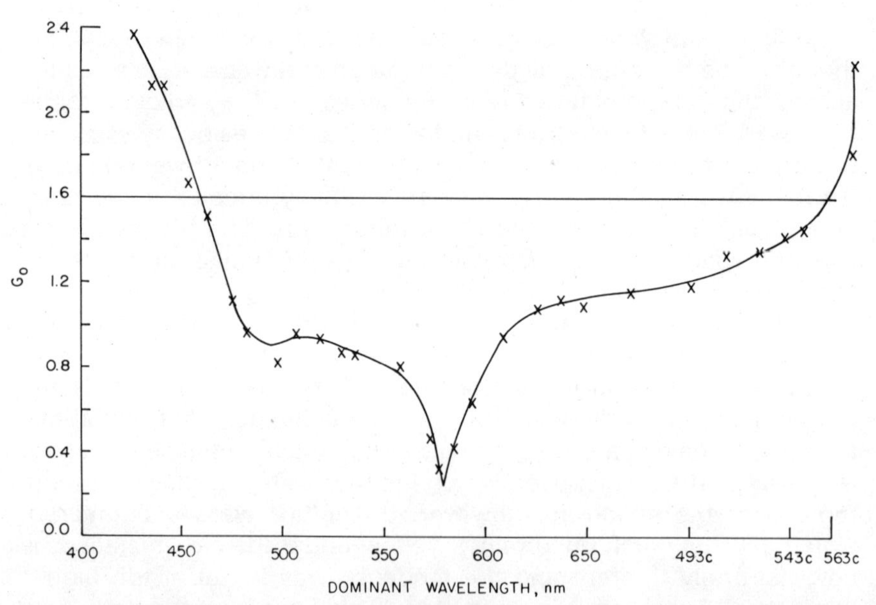

Fig. 7-1. G_0 as function of stimulus monochromatic wavelength with achromatic (C. 7000 K) surround at 100 mL (corrected for stimulus purities; observer RME). (From *J. Opt. Soc.* **59,** 633, 1969).

The figure, therefore, is in terms of the "density" of the stimulus relative to the surround as zero. Formally, in terms of the luminances it is log L_s/L_b where L_s is the luminance of the surround and L_b is the luminance of the central stimulus (main beam). For those familiar with photographic densitometry this is the reflection density of the stimulus considered as on a white support. It turned out that this was a fortunate choice because it enabled us to recognize the form of the function as similar to that of a well known function found in an entirely different manner but usually expressed in this rather peculiar, logarithmic-type ordinate from sheer convenience of plotting. This is the so-called "purity threshold" function for the first hue perceptible when monochromatic light is added to an achromatic stimulus. This function is usually plotted as log $1/p_c$. It was, in fact, largely the later proof of the identity of these functions that served as the key to the problem of its meaning.

This function and others closely related to it have appeared in the literature a number of times, and the widely different approaches by which they have been derived are sufficiently indicative of its fundamental nature. It is worthwhile to review some of them briefly; I shall not attempt to do it historically.

THE PURITY THRESHOLD

The relative luminances required when two complementary wavelengths neutralize each other to form an achromatic mixture vary widely with the wavelengths involved, as do the luminances for three-part mixtures. Since the time of Helmholtz this has been interpreted as meaning that the different wavelengths have different "saturations," that is, produce differing degrees of achromatic perception in addition to the hue. There have been a number of studies made to determine this relative "saturation." Good reviews are given by Graham (1965) and Wyszecki and Stiles (1967), as well as LeGrand (1957).

The primary method employed has been to determine the amount of monochromatic light that must be added to white light to produce a just-perceptible hue. Since colorimetric purity is the ratio of monochromatic to total luminance in a color, this luminance divided by the sum of the two is the colorimetric purity (p_c) threshold for the wavelength measured. The function so found, when plotted as the logarithm of $1/p_c$ against wavelength gives a curve essentially identical in shape, considering the variability of such determinations, to our G_0 curve. They differ in the value of the ordinates but are everywhere essentially parallel. It follows that G_0 and log $1/p_c$ differ by a constant (logarithmic) amount or, expressed linearly, that the two variables are related by a single multiplying constant. The two variables are, however, somewhat different in nature: colorimetric purity is the fraction of monochromatic light in a mixture ($1/p_c$ is, of course, the inverse of this), and the threshold at G_0 is the ratio of the surround (white) luminance to that of the monochromatic central stimulus. Thus one is the ratio in a mixture and the other is a ratio of two adjacent luminances. We return to this in some detail later and show that the two are, in fact, simply connected. The point needs to be made here is that the simple explanation of this function as caused by differences in "saturation" of the spectral colors, while plausible for the purity threshold, is not equally applicable to the G_0 threshold. What is logical is that both are indicative of what I shall call the varying "chromatic strengths" of the different wavelengths.

MUNSELL CHROMA

The Munsell system is an orderly arrangement of reflecting color samples intended to produce perceptions uniformly spaced along the three dimensions of Hue, Value, and Chroma (capitalized to imply their definitions). I shall discuss the whole system in some detail in Chapter 9. Chroma is defined in this sytem as the amount of the difference of a color from a gray (achromatic) of the same Value (luminous reflectance). Chroma is spaced uniformly on an arbitrary scale from the gray as zero out to the maximum attainable pigment purity. Colors of constant Value and Chroma thus form a closed group of colors that differ only in hue. The assignment of samples to such a group (in fact the whole arrangement) was initially done entirely visually without reference to colorimetry.

An early study of such a visually produced "Chroma circle" containing 100 Hues was made by Nickerson and Granville (1940). In this study they assumed Chroma, which we shall find to be a complex variable, to be equivalent of saturation, and they wanted to show its relation to colorimetric purity. To do this they determined the dominant wavelength and purity for each sample by direct spectrophotometry. They showed that the function so obtained correlated very well with the results obtained by the purity-threshold technique we have just considered. Their conclusion, "It seems obvious that the relation is considerably more than a casual one," is a considerable understatement. We have confirmed this result for more recent samples and have also found that a constant value of G_0 is probably the requirement for the chromaticities of such a circle. The fact that the selection of their samples was purely visual and based only on the "difference from a gray" again supports the concept of chromatic strength as the variable involved, this time at much higher purity levels.

HERING OPPONENT HUES

Hering, working somewhat later than Helmholtz, pointed out that the unique hues, red, yellow, green, and blue, cancel each other in pairs (see Hering 1905-1920). It is not possible to have a stimulus that produces *both* red and green perceptions, nor blue and yellow. Thus a yellowish-green and a reddish-yellow, if mixed optically, can be so adjusted in ratio that only yellow remains; a yellowish green and a bluish green mixture can be so adjusted that only green remains; and so on for all possible pairs of different colors. On this basis Hering proposed a theory of color vision involving receptors with three opponent processes that responded to either green *or* red, blue *or* yellow, and black *or* white.

Hurvich and Jameson undertook to investigate this theory quantitatively and in a series of papers (see Hurvich in the bibliography) showed that on this basis many of the known psychophysical relationships can be predicted. Our concern is the fact that among these predictable relationships is the purity threshold curve. To understand how they arrived at this result it is necessary to consider their approach in some detail.

The unique hues are defined as a yellow that is neither greenish nor reddish, a green that is neither yellowish nor bluish, and so on for blue and red (which is extra-spectral and has to be a mixture of long- and short-wavelength stimuli). The exact wavelengths for each differ somewhat between observers but are not difficult to recognize. Having determined their own unique stimuli, Hurvich and Jameson then determined for each monochromatic wavelength the amount of the appropriate opponent stimulus that would reduce each to a unique hue by admixture. Thus for a wavelength seen as a bluish-green, for example, they determined the amount of yellow it was necessary to add to make the mixture a unique green, the mixture being determined by *recognition* of the unique hue. Each wavelength that did not itself produce a unique hue could obviously be modified in two ways to produce one. The bluish-green above could be modified to unique blue by the addition of unique red or to unique green by the addition of yellow, and so on for all wavelengths. They thus obtained, for all wavelengths (at constant mixed luminance), the amounts of the two unique hues necessary to cancel the two hue components of the color produced by each wavelength. This results in two unrelated sets of data, one for blue-yellow components and one for red-green. In order to relate these to each other they determined, again by recognition, the wavelengths that produced an equal-part mixture of the adjacent unique hues; red-yellow, yellow-green, green-blue, and blue-red. They then defined the "color valences" of the unique hues as equal to each other at these points and adjusted their data so that both sets were expressed in the same units. Each wavelength could then be described as producing two hues in amounts specified in terms of the amounts of their opponent colors necessary to neutralize them. The bluish-green, for example, could be expressed as a (relative) quantitative amount of blue and green excitation. The sum of these two for any wavelength is thus a measure of the total *chromatic* excitation produced by that wavelength.

As a separate study Hurvich and Jameson had each determined his own spectral luminous efficiency curves. Reasoning (from the Hering theory) that saturation would be indicated by the ratio of chromatic to achromatic excitation and that the purity threshold also indicated saturation, they carried out this computation and found that the two curves were, in fact, parallel when plotted in logarithmic units; the logarithm of the ratio of their luminance to chromatic valence at each wavelength gave a curve parallel to the log $1/p_c$ curve from the literature.

We shall return to these studies later but should note at this point that they are unique in that the perceptual phenomena on which the results are based do not involve a match between aspects of two stimuli seen simultaneously. The chromatic valences are based on cancellations of hue components with the end point a recognizable (unique) hue. The results were obtained, however, with an achromatic surround of constant luminance. They are thus based wholly on perceived *hue* without regard to any of the other perceptual variables. The concept of "saturation" of the stimuli is hence irrelevant to these results, and the concept of chromatic strength, at least for cancellation with respect to each other, is almost inevitable.

COMPLEMENTARY STIMULI

The cancellation of complementary stimuli to produce a white has been known since Newton, and the similar cancellation of three stimuli is the basis of tristimulus colorimetry. It has also been well known that the relative luminances of the different wavelengths producing these cancellations differed widely. In tristimulus colorimetry these differences are introduced into the units in which the primaries are represented by arbitrarily equating the amounts that form this white. Thus in the CIE System, $X = Y = Z$ is the condition for a match to the equal-energy achromatic point.

The luminance relations involved in *pairs* of complementary stimuli were the subject of an early study by Sinden (1923). In this study he determined the ratio of the luminances required to match an arbitrary white (c. 5000 K). He attempted by a number of assumptions to place these values on a comparable basis and obtained a curve intended to show the relative saturations of the various wavelengths. This function bears a similarity to our G_0 results sufficient to justify the assumption that the same visual process is involved; the minimum is quite sharp at 575 nm (which he arbitrarily set at unity), the general shape of the curve is similar, and the maximum in the violet approaches a factor of 100 as does ours. The major difference lies in the green region which is much lower than ours but which he only approximated.

MACADAM MOMENTS

Some years later, MacAdam (1938) showed that results similar to those of Sinden could be derived directly from the CIE chromaticity diagram, since the luminance ratios of the complementary stimuli for a given white point are inherent in the diagram. Reverting to the Newtonian concept of mechanical moment about the white point, he defined the moment of any chromaticity as its mass $(X + Y + Z)$ multiplied by its distance from the white point on the CIE diagram. He then pointed out that the moments of any two complementary chromaticities stand in inverse relation to those of the luminances required to form the white in combination. For spectral stimuli they are thus the ratios of Sinden's raw data. By assigning a value of unity to each wavelength in turn he thus obtained a figure for the moment of the monochromatic stimuli in terms of their complementary stimuli, and, by extension, those of all other chromaticities on the same basis. This is the basis of the chromatic-moment diagram given in Chapter 4.

The plotting of this function for the monochromatic stimuli against wavelength results in a curve superficially like that of Sinden's but that differs significantly both in shape and in the position of the minimum. This is not particularly apparent from Sinden's data but is readily apparent if it is compared instead to our G_0 function. We consider this difference later in the chapter. MacAdam showed that the ratio of the moment of a given chromaticity to that of its complementary wavelength is the colorimetric purity of that chromaticity, a fact we shall also consider later.

We see that four very different approaches lead to the suggestion of a stimulus characteristic that varies with wavelength in the same manner as does the G_0 threshold. In the case of the fifth approach the results are suggestively similar but with a significant difference. We proceed, therefore, to characterize this concept of chromatic strength as derived from our work, and consider it in relation to these studies as well as to other known psychophysical relationships.

BRILLIANCE

We need first to consider in more detail a concept mentioned earlier, that of "brilliance." The perceptions of grayness and fluorence are quite different from each other, and they appear at first glance to have little in common. Experimentally, however, they are mutually exclusive; it is not possible to produce a stimulus in which both are seen simultaneously, and the transition from one to the other is always by way of the G_0 threshold. Furthermore, both depend on the characteristics of the surround stimulus. It was apparent fairly early in the work that the threshold must in some sense represent a situation in which some perception from the stimulus matched an equivalent perception from the surround. We found, eventually, that "brilliance" was an excellent term to describe this common appearance. In fact we found that completely naive observers were often confused when we talked about "grayness" and "apparent fluorescence" but could set G_0 without hesitation when we asked them to set a "brilliance match."

There are a number of possible objections to the use of the word "brilliance" for this factor. In the 1921 report on colorimetry by the OSA. committee under Troland, "brilliance" was used in place of "brightness" because the Illuminating Engineering Society had adopted a conflicting meaning for the latter. In the revision of that work under Jones, published as the *Science of Color* in 1953, this objection no longer held and "brightness" was again used. "Brilliance" is also widely used in industry, particularly in relation to dyestuffs, in specific meanings not obviously related to the concept meant here; "brilliant black" is an example. Nevertheless I shall adopt the word for present purposes and I think we shall find that, in fact, the intended concept is closer to common speech usage than at first appears. I shall speak of a "brilliance match" when a stimulus is at G_0 with respect to its surround and then, later, generalize the meaning to the perception of any two or more stimuli that under the conditions show the *same degree* of grayness or fluorence. "Brilliance," therefore, will be used as the name of a separate, continuous, perceptual variable that is seen as less than that of the reference stimulus when grayness is present and more when fluorence is seen. We can now quantify the concept of chromatic strength in terms of this single perceptual variable.

CHROMATIC STRENGTH

If we return to the situation of a monochromatic stimulus seen against a much larger, achromatic, surround we can say that the condition necessary for a brilliance match between the two is that the luminance (L_b) of the central stimulus be below that of the surround (L_s) by a factor that is characteristic of the wavelength or, inversely, that the luminance of the stimulus would have to be multiplied by such a characteristic factor to equal that of the surround. I shall call this latter factor the chromatic strength (S). Since $G_0 = \log L_s/L_b$ we can thus define S as equal to $\log^{-1} G_0$. For monochromatic light it can be stated as $S\lambda$. Since G_0 can be set for any chromaticity we can generalize this to $S_{x,y}$ where the subscript refers to the chromaticity (x, y). For the general case of any chromaticity at any luminance against a specific achromatic surround, the characteristic of the stimulus that produces the perception of brilliance relative to the surround can then be stated as $S_{x,y} \times L_{x,y}$ and evaluated with respect to the surround luminance, considered to have an S equal to 1. We shall see later that this can be further generalized, at least in concept, to chromatic surrounds. It is important in what follows to remember that while S_λ and $S_{x,y}$ can be considered constants for monochromatic light or any other chromaticity with respect to a given surround, they are *not* characteristics of the dominant wavelength or the indicated chromaticity *per se.* They are characteristic of the potential of that stimulus to evoke brilliance *against that surround.* In other words, brilliance must be thought of as the perception of the state of excitation of the eye produced by the stimulus *relative* to that produced by the surround and S as a measure of this. It is just as susceptible to a change in the surround as in the stimulus.

For any chromaticity (x, y) we can, of course, substitute its metameric "monochromatic plus white" equivalent. For this stimulus it is not unreasonable to assume that its brilliance will be the sum of the brilliances of each component, with the monochromatic component having the chromatic strength S_λ and the achromatic component (assumed the same chromaticity as the surround) having $S_w = 1$. These assumptions permit us to write an equation for brilliance in terms of colorimetric purity and dominant wavelength that can be tested experimentally; that is, for any chromaticity the sum of the products of the luminance of each component by its appropriate S should equal the luminance of the surround ($S_w = 1$) at the G0 threshold.

In setting up such an equation it is necessary to take into account the fact that the luminance of the chromatic component must attain an appreciable level before it is seen as a hue component of the mixture; this is the purity threshold and is of considerable magnitude for some wavelengths. We assume that S_λ applies only to the luminance above this threshold but that the part below contributes to the total luminance. The threshold need not be considered for high purities where most of the luminance is due to the chromatic component itself.

There are five luminances (L) involved in the complete equation: L_s that of the surround, L_b that of the total stimulus, L_w that of its white component, L_λ that of the monochromatic component, and $L_{\lambda T}$ that of its threshold. And there are two chromatic strengths: S_w that of the white (assumed $= 1$), and S_λ that of the pure monochromatic component. At G0 this gives:

$$L_s = L_w + L_{\lambda T} + S_\lambda\ (L_\lambda - L_{\lambda T})$$

but $p_c = L_\lambda/(L_\lambda + L_w)$ and $L_b = L_\lambda + L_w$, so that if we neglect $L_{\lambda T}$, we can write $L_s = L_b\ [(1 - p_c) + S_\lambda\ p_c]$. Furthermore, since, at G0, $L_s/L_b = S_{x,y}$ we can write

$$S_{x,y} = 1 + p_c\ (S_\lambda - 1)$$

which should hold for high-purity stimuli.

Since all of our filters gave high purity stimuli this equation was used to extrapolate our results for our G0 values to a purity of 1, and it is these results that were used for the plot of Figure 7-1.

THE CHROMATIC THRESHOLD

In order to predict the G0s for a complete purity series and thus be able to check our assumptions experimentally we need to know the values of $L_{\lambda T}$. As we noted earlier, the parallelism that we found from the literature between G0 and the values of log $1/p_c$ suggested that the two are connected by a constant factor. They represent, however, quite different perceptual phenomena because G0 is a threshold for a color *against* white and the p_c threshold is for the color in a mixture *with* white.

The actual relationship between the two was found in a series of studies that turned up a fact which is apparently new to the literature. We found that, with a monochromatic stimulus in the central beam, when the luminance was reduced until hue disappeared (reduced to the "black point"), the value of the luminance was a constant fraction of the luminance of the G0 setting for that stimulus, independent of wavelength. We thus established that the black point curve also paralelled the log $1/p_c$ curve. The black point, however, can also be considered as the hue threshold for the stimulus against white, and we called it the "chromatic threshold." We then discovered the surprising fact that if this monochromatic luminance was left constant and white light was added, the mixture remained at the chromatic threshold all the way up to a luminance match with the white surround. That is, for any admixture of white, the slightest increase in the chromatic component made a hue visible. The relationship between the black point and the "purity threshold" is thus an identity, and "purity threshold" is a misnomer since the threshold is entirely independent of purity. The luminance of the chromatic threshold ($L_{\lambda T}$) is thus a constant fraction of the luminance of G_0 for any wavelength.

This study, like so many others in our work, demonstrates the almost complete independence of the chromatic and achromatic aspects of the perception of a color stimulus. It means that the monochromatic plus white analysis (and hence p_c) of a stimulus is a direct key to the perceptions it invokes. It is on this basis that I shall presently define "saturation" as a perception based on the relative brilliances of these two components.

We found that, although there was a difference between the chromatic thresholds of our two observers, for a single observer there was an almost perfect correlation between the values predicted from the complete equation and the experimental results. The equation is thus justified and with it the assumptions: (1) that $S_w = 1$ for both the surround and the achromatic component of the stimulus, (2) that $S_\lambda L_\lambda$ gives the brilliance of the chromatic component, and (3) that the brilliance of the stimulus against the surround is given by the sum of the separate brilliances. It is apparent that the foundations are thus laid for calculations for luminances comparable to the surround based on chromatic strength, given the values for monochromatic light; that is, these results are based on brilliance matches to the surround.

THE ACHROMATIC THRESHOLD

In a stimulus with a purity of less than 1 there must also be a threshold for the achromatic component that is a constant for the conditions and the observer. Direct determination of this threshold by the two of us (at widely different times) gave distinctly different results. Thus BS found a value of 1.6 ($^1/_{40}$ the luminance) and RME a value of 1.0 ($^1/_{10}$), perhaps as a result of the cataract condition. The value even for BS, however, is at a distinctly *higher* luminance than that of the chromatic threshold for any wavelength. This somewhat unexpected result required confirmation and led to a rather startling demonstration of its truth. After careful determination of the point for a stimulus exactly matching the chromaticity of the surround, each filter was in turn substituted in the beam with its luminance adjusted to this value. In *all* cases hue was clearly perceptible and for the blue region was actually fluorent. The actual appearance of the stimuli is indicated directly by the curve of Figure 7-1 if a line is drawn parallel to the wavelength scale at 1.6. All wavelengths having G_0's above the line are fluorent, the intersections are at G_0, and those below have varying degrees of grayness indicated by the distance, the yellow approaching black. Thus only the yellow (c. 575 nm) has a chromatic threshold luminance comparable to the luminance threshold of white; all others are lower.

THE GENERALIZED BRILLIANCE EQUATION

In order to state a general equation for the brilliance of a stimulus with a purity of less than one *and* a luminance *different* from that of its G_0, it is thus necessary to include a term (L_{wT}) for the achromatic threshold. The complete equation for a brilliance match then becomes:

$$S_w L_s = S_w(L_w - L_{wT}) + S_\lambda(L_\lambda - L_{\lambda T}) + L_{wT} + L_{\lambda T}$$

where $S_w = 1$, $S\lambda$-$\log^{-1} G_0$ for light of wavelength λ

L_w and L_λ = the luminances of the white and monochromatic components

L_{wT} and $L_{\lambda T}$ = the respective thresholds against the surround

L_s = the luminance of the surround

This equation holds specifically only for a brilliance match to the surround, thus making L_w and L_λ related variables; given the value of one the other is determined. Considering the independence of the two perceptions that are involved, it is not unreasonable to assume that the right-hand side of the equation represents the brilliance of any stimulus of chromaticity (x, y) and luminance $L_{(x,y)}$, that is, we may substitute $S_{(x,y)}L_{(x,y)}$ for $S_w L_s$. Justification for such an assumption comes from an early study in the same series (Evans and Swenholt 1968). In this study a small strip of neutral density was introduced into the surround field and conditions were determined at which stimuli showed the same grayness as the strip. For all wavelengths the luminance for this condition was a constant log luminance decrease below the corresponding G_0 value. Thus we assume that, as seen against an achromatic surround, any two stimuli that calculate to the same value will have the same brilliance whether this be seen as grayness, fluorence or a brilliance match to the surround.

On the basis of this assumption we can write:

$$S_{(x,y)}L_{(x,y)} = [S_w (L_w - L_{wT}) + L_{wT}] + [S_\lambda (L_\lambda - L_{\lambda T}) + L_{\lambda T}]$$

where the brackets indicate the two components of the monochromatic plus white mixture metameric with the chromaticity (x, y). As applied to our data in which the surround was at 100 mL, this means that values above 100 for this equation indicate fluorence, those below indicate grayness, and black occurs when both $L_w = L_{wT}$ and $L_\lambda = L_{\lambda T}$ or when both are lower than these values. It is apparent that if either L_w or L_λ is set equal to 0 and the other permitted to vary, this equation describes the perceptual series for monochromatic and achromatic stimuli as discussed earlier. The equation does not, of course, quantitatively describe the series so generated; the perceptual scale is more nearly the logarithm or the cube root of the brilliance number so calculated. It does, however, correctly describe the brilliance of a stimulus with respect to either the surround or another stimulus and with regard to its fluorence or grayness, in the sense of greater or less. The equation is also not meaningful, as we have seen, when the total luminance of the stimulus ($L_w + L_\lambda$) considerably exceeds that of the surround.

Of particular interest is the case of a stimulus of intermediate purity in which both chromatic and achromatic components are present. We want to consider the perceptions generated by such a stimulus as its luminance is varied with respect to the achromatic surround and take as an example a stimulus with $p_c = 0.5$; that is, $L_w = L_\lambda$ for the whole series. Suppose we take a wavelength (from the green or red) for which $S_\lambda = 20$ ($G_0 = 1.3$) and assume that $L_{\lambda T} = 0.25$ (1.3 below G_0) and $L_{wT} = 2.5$ (1.6 below the surround).

If we start with a stimulus luminance of, say, 0.1 mL ($^1/_{1000}$ of the 100 mL surround) black will be seen, and there will be no change in this black as the luminance is increased until $L_\lambda = L_{\lambda T}$, that is, at a total luminance of 0.5 mL. At this point the chromatic threshold is reached, and slightly above it a hue is seen in the blackness. As luminance continues to rise this hue becomes stronger and blackness decreases somewhat until a value of 5 mL is reached, which is the threshold for the achromatic component. It is not known whether or not this is the point at which the blackness perception is more appropriately described as grayness; it is an interesting point which needs investigation. In any case, as luminance is further increased grayness decreases (brilliance increases) until the value of the equation becomes 100. This occurs when the stimulus luminance is slightly less than 10 mL (9.78), and from this luminance on up to a luminance match fluorence appears and increases steadily. When the luminance becomes of the order of twice that of the surround this perception rather suddenly disappears, as discussed earlier.

We see then that in a mixed stimulus with purity less than 1, the perception of the achromatic component occurs at a considerably higher luminance than the first perception of hue, that grayness and blackness are associated with the brilliance and not only with the achromatic component, and that both components together determine the point at which grayness disappears and fluorence starts. These points have important bearing on the perception of saturation.

While the above discussion is valid for our instrument and can occur in everyday situations, it is not directly applicable to real reflection colors because of the unavoidable achromatic surface reflection of such colors. To explain briefly, if we set this surface reflectance at 2½%, the minimum value of L_w is already equal to L_{wT}. This means that achromatic perception is present even when there is no reflection from below the surface and the low-reflectance end of the series is modified accordingly. Actual computation would depend on whether the surface reflection was included in the purity calculation, but the upper end of the series, in the neighborhood of G_0 and a luminance match, would not be affected, assuming such levels could in fact even be reached with real pigments. They can be, and often are, reached with real pigments if the sample is separately illuminated from the comparison background, and the case is thus of real, practical, importance. I shall consider such situations in later chapters.

The above discussion has been limited to the perception of the brilliance evoked by a stimulus seen against a white surround. The variables L_w and L_λ, however, completely define such a stimulus, and it is worthwhile at this point to review the other four variables in these same terms before going on to consider brilliance in terms of the CIE variables.

LIGHTNESS

Lightness appears to be the perception of relative luminance, in our case stimulus luminance $(L_w + L_\lambda)$ relative to that of the surround (L_s).

It is thus independent of S_λ. It is conventionally considered as being 0 when the luminance is below the threshold and as having a value of 1.0 at a luminance match. The evidence that lightness is in fact due to the luminance ratio comes largely from the fact that the luminance match point can be set visually by an experienced observer for all stimuli. For high purity stimuli this occurs in a region of high fluorence and hence demonstrates that relative luminance as such is a directly perceived variable.

There is some dispute as to whether or not there may be a systematic difference between a perceived and a calculated luminance match, especially at high purities, when the stimulus is seen against a white surround; that is, whether heterochromatic brightness matching to the surround is exact under these conditions. The literature is not interpretable on this point because brilliance has not been isolated as a variable, but, in any case, the reported discrepancies are small compared to the difference between the match point and G_0.

Against a constant luminance surround seen as white, lightness and brilliance are associated variables, both being determined by the values of L_w and L_λ. The relationship between them is, however, a steep function of both wavelength and purity because of the wide range through the spectrum in value of S_λ that affects only the brilliance. Thus two stimuli of the same lightness and purity but different hues may differ to the extent that one is flourent and the other quite grayish.

In real reflecting samples the achromatic surface reflectance is seldom less than 2½% and in the absence of a sufficient chromatic stimulus from below the surface is normally seen as a black against a white surround. Since this is also the achromatic threshold, all reflection stimuli have an achromatic component unless this reflection is eliminated. The maximum theoretical reflectance at any wavelength is, of course, 100 so that any reflection stimulus whose color is due only to absorption (that is, any non-fluorescent stimulus) must have a luminance and hence lightness very considerably less than the surround. For this reason few reflection colors can reach, even theoretically, a lightness at which the brilliance is equal to the surround, and most reflection colors have a relatively large amount of grayness, depending on their hue. When such a color is separately illuminated by a higher intensity than that of the surround then the rapid increase in its brilliance (because of its S_λ factor) as compared to its increase in lightness becomes at once apparent.

Nevertheless, for a given background, brilliance and lightness are related and, for a given chromaticity and relative luminance, one can be predicted from the other. We shall see in the next chapter that when the surround is also chromatic they become independent because of the dependence of S_λ on surround chromaticity.

SATURATION

The third perceptual variable, saturation, is more difficult to treat because of the almost complete ambiguity of the word as used in the literature, apparently starting with Helmholtz. Even in the nontechnical literature the belief in only three perceptual color variables (which predates Helmholtz) produced the same conflict in concepts and he may be said to have only supplied the word. One meaning of the word is universally accepted, that of the perception of an achromatic component in a color. The difficulty is that it has also been used throughout to include what we are separating out as brilliance.

Earlier, I described saturation as the perception of "effective" colorimetric purity. In essence this likens the concept of saturation to that of hue mixtures; that is, the chromatic and achromatic components of a color are seen separately and their relative amounts judged in the same way that red and yellow can be separately judged in mixtures from red through orange to yellow, and it is in this sense that I shall use the word. The concepts are somewhat different in that we tend to think of chromatic and achromatic responses, but not of yellow and red responses, as different *in kind*. We think of the hue as occurring *in* the white or vice versa; we do not think of the yellow as being *in* the red. This is probably due to the existence of p_c colorimetric purity as a variable and the absence of a corresponding variable for hue. Because of this, and because of the obvious analogy with the present day use of the word in chemistry, we think of saturation as relating to the portion of the total color perception that is chromatic.

Incidentally, at the time of Helmholtz, the verb "saturate" had the meaning in chemistry that today we associate with "neutralize," that is, to produce a salt by a mixture of an acid and a base until neither is in excess. The analogy of this with the cancellation of complementaries to produce a white may have been one reason for the confusion since both involve the concept of the relative strengths of two components. Our use of the word is more analogous to the concept of the concentration of a colored compound in a colorless solvent although this should not be pursued very far either.

In any case, the concept of saturation as it pertains to the relationship of hue perception to the total color lines it up directly with the psychophysical concept of colorimetric purity. Proceeding to consider this relationship, we encounter at once the fact that two colors of different dominant wavelength but the same purity can have widely different degrees of apparent admixture with white. The usual explanation of this has been that the various spectral colors themselves produce differing amounts of achromatic response. This explanation overlooks the magnitude of the saturation differences that would be required in the spectral colors and the simple fact that, aside perhaps from yellow, no achromatic component at all is usually perceived in a pure spectrum. To assume instead that the various monochromatic stimuli differ in chromatic strength, that is, brilliance, accords far more nearly with what is actually seen.

I shall assume, therefore, that saturation is the perception of the brilliance of the hue component of a color in relation to its total brilliance. This corresponds to the ratio of $S_{\lambda}\ (L_{\lambda}-L_{\lambda T})$ to the total $S_{(x,\ y)}L_{(x,\ y)}$. This is what was intended above by the term "effective purity" and suggests the term "brilliance purity" as more closely allied to the perception. Since colorimetric purity is given by $L_{\lambda}/L_{(x,\ y)}$, it is apparent that brilliance purity varies with colorimetric purity but at a different rate for different dominant wavelengths depending on S_{λ}. When thresholds are taken into account, however, the relationship is not simply that of a multiplying factor.

It follows from this definition that saturation varies from zero when no chromatic component is visible, up to a maximum when no achromatic component is visible. This differs from the usual view only in that it postulates the existence of a maximum that may occur, for some wavelengths, *before* a purity of one is reached; the usual statement is that this does not occur for *any* wavelength. Under our assumption the maximum may be reached at relatively low colorimetric purities.

For example, in the extreme cases, the brilliance due to monochromatic blue may be upward of 200 times that of the white component per millilambert. The white may thus become imperceptible considerably before a purity of one is reached. For a yellow, on the other hand, the monochromatic may exceed the achromatic by a factor of only 2 to 4 per millilambert, and so the achromatic may remain perceptible to very high colorimetric purity values. If, as many if not most writers imply, monochromatic yellow also produces an achromatic perceptual component, then these views would have to be modified accordingly and maximum saturation would not be reached at a purity of 1. This seems to me an unnecessary assumption at this point. While there is very much evidence that the perception of yellow closely resembles that of an achromatic stimulus, its low chromatic strength would appear to account for this satisfactorily without postulating an achromatic component.

To the extent that these statements are true, they mean that *all* statements in the literature that claim that such and such a situation changes the perceived saturation must be reexamined to determine whether saturation or brilliance was, in fact, the perceptual variable that changed. The usual statement, for example, that the "saturation" produced by monochromatic light can be greatly increased by preadaptation to its complementary, while possibly true for some wavelengths, seems to me more likely to refer to the chromatic brilliance seen, perhaps even to fluorence.

If we combine our definition of saturation with the assumption on which it is based, that the chromatic and achromatic components are separately perceptible, then two other facts can be noted which are important to later discussions.

It is one of the tenets of the workers who base color-order systems on the Hering theory, notably Johannsen and Hesselgren in Sweden (*The Natural Color System*), that an observer is always conscious of the appearance of a pure hue for any dominant wavelength, that is, of its appearance uncontaminated by white. This follows directly from our assumptions but would be essentially impossible under the assumption that no stimulus can produce a saturation maximum.

Under our assumptions and definition it would also follow that if we start with an isolated stimulus ($S_\lambda = 1$) having a purity of one and gradually add an achromatic stimulus to it, the luminance threshold for its perception should be found to be independent of the initial wavelength. Wright and Pitt (1935) found this to be so, and it was suggested that the unexpected result was somehow due to adaptation. It is difficult to imagine how such an adaptation would operate if the different wavelengths produce different degrees of achromatic response. Under the assumptions that the perceptions are entirely separate and that no wavelength produces any achromatic response, it is obviously the result to be expected. The experiment becomes a crucial one and needs careful reinvestigation, especially in the neighborhood of stimuli seen as yellow. It does not follow from our treatment that the saturations will be alike at these thresholds because of the variability of S_λ. Actually we shall find in the next chapter that under these conditions it is always true that $S_\lambda = 1$, so that, *in an isolated stimulus,* the saturation maximum should be reached at the same colorimetric purity for all wavelengths, but not in a stimulus seen against white.

HUE

The fourth perceptual variable we need to discuss here is hue. Of all the phases of color perception, hue has received by far the least scientific attention. At first glance this seems surprising because hue is the most obvious perceptual variable of color, but it is most unsatisfactory as a variable because there is no obvious way in which it can be quantified. Furthermore it is a most unstable variable, dependent on almost all of the parameters of any physical set-up that can be arranged for its investigation. In general, it is not possible to have a comparison stimulus present simultaneously with that being studied without the likelihood that the standard will be changed by the presence of the second stimulus. The investigator is thus thrown back on short- or long-term memory matches, both of which are well known to be systematically poor. The alternative to these approaches is inter-eye comparison methods but these have been considered respectable only in quite recent years. I am sure that a systematic study of perceived hue by this technique would unravel most of the mysteries of color perception. My certainty comes from my very strong feeling that only a complete understanding of hue and brilliance as variables will tell us what kind of phenomenon it is for which we are trying to postulate a mechanism. It has long been assumed that all the facts of color perception can be deduced from knowledge based on the fact of metamerism alone, that given a sufficient elaboration of these facts all perceptions can be deduced. This appears not to be the case, and I feel that hue and its *associated* variable brilliance may well be the missing links. I shall return to this matter when we consider the relation of brilliance and hue to the CIE system, immediately following the discussion of brightness below.

BRIGHTNESS

The fifth perception that we noted earlier, that of brightness, is of course present in all these series but is not, so to speak, a relevant variable. We shall see that, except in the case of isolated stimuli or those much brighter than any others present, the concept refers to the perception of the total actual or relative light intensity of the *whole field.* Thus in our instrument and all similar situations, it refers essentially to the perceived apparent intensity of the surround itself.

An experiment not yet referred to is relevant to this point as well as to the general validity of the G_0 function. In one, rather carefully done series of studies (Evans and Swenholt 1969), the luminance level of the surround was changed from less than 50 mL to greater than 800 mL. We concluded that over the range from about 50 to 500 mL the G_0 curve was invariant with respect to the surround. No further work was done on this, no black point or purity series were run, but we became convinced that this invariance could be generalized to all situations, that is, that G_0 was strictly relative to the luminance that controlled the sensitivity of the eye. For this reason all further work was done at a constant level of 100 mL.

We might note here that this finding is in apparent contradiction to common experience (and a good deal of the literature) in that distinct differences are found in related fields when the overall luminance is changed. While we did not investigate the point we feel reasonably certain that these differences are due primarily to changes in the chromatic and achromatic thresholds of the stimuli, thus changing all colors in which these play an active role.

The reasons for choosing the 100 mL level are pertinent to the present discussion. When the level was set below 50 mL, and even at that level, the settings for G_0 for some wavelengths were falling in a luminance range below the photopic level and hence not only difficult to make but suspect on many grounds. Above 500 mL, particularly at 800 mL, two related phenomena occurred. The surround appeared dazzlingly bright, and any attempt to set G_0 was not only uncomfortable but correspondingly uncertain. Brightness, then, for our instrumental case, was represented by the appearance of the whole field, dim and uncertain below 50 mL, dazzlingly bright at 800 mL, and constant for any given luminance.

Quite aside from the illustration of how brightness perception entered into our work, two things should be noted here to which we shall have occasion to refer later. The first is that brightness is a separate, independent perception, not directly affecting any of the others. For this reason, only confusion results from any attempt to use the word in connection with different *parts* of the field. It is true, of course, that each part of the field is seen with its own brightness, and it is legitimate to say that one part of the field is brighter than another. In any complex scene, however, these facts are completely obscured by the other perceptual variables insofar as *individual* stimulus areas are concerned. Brightness is the perception of the aggregate level of all the stimuli in the area to which it refers. Only in the case of an isolated stimulus does it become a perception due to a *single* stimulus. We shall see in a later chapter that its most appropriate use is in reference to the apparent *illumination* level of a scene or an area in that scene. In such cases it becomes an obvious, independent, perceptual variable.

The second important point from our instrument study is that a 10° field, with no stimulus outside of it, is not sufficient to control the sensitivity of the eye, except for stimuli that it completely surrounds. We here border on a subject we will discuss later (the effect of the adjacency of two stimuli), but the fact that 800 mL appeared dazzlingly bright means that the eye did not completely adjust to that level and that this was *caused by* the fact that the field was only 10° across. A daylight illumination of 800 mL borders on being a dark day! In the next chapter we shall be considering chromatic surrounds at 100 mL and will find some of those uncomfortably bright. We have to think of brightness, then, in terms of all visible stimuli in the field. While we feel that it is legitimate, on the basis of the invariance of G_0 and many other facts, to consider the sensitivity level of the eye for the surrounded stimulus as set by the surround, we feel that it is not legitimate to assume this for the surround itself. If we reserve "brightness" as a description of this latter perception then we not only are less likely to confuse it with brilliance and lightness but will also be in a far better position to consider the truly complex fields of everyday life.

CHROMATIC STRENGTH AND THE CIE SYSTEM

We noted earlier that from the start of our investigations we had expected that accurate determination of the variables affecting our G_0 threshold would permit its derivation from the CIE data and hence establish it in terms of the Standard Observer. It now seems likely that the trichromatic mixture functions and the relative efficiency data are not sufficient to determine this function and that added psychophysical relations are necessary. It is, however, obviously true that any such relations must be consistent with any function derivable from CIE, since there is no doubt that the CIE system specifies all possible stimuli. The difficulty seems to lie in the fact that brilliance is the result of interaction between unmatched stimuli whereas the system deals only with metamers. On the other hand the CIE system comes tantalizingly close to predicting actual appearance in many situations, and it is my impression that the system is simply distorted rather than inadequate.

It would take us far outside the scope of the present work to attempt to resolve these problems here, but it is worthwhile to sketch out the nature of the necessary relationships.

We have already discussed the similarity of the "MacAdam moments" for the spectral colors to the results of Sinden and to our own results on the G_0 threshold. We found early in our studies (Evans and Swenholt 1967) that these moments could in fact be derived from our data. The ratio of the moments of complementary monochromatic stimuli is identical, well within experimental error, to the ratio of the luminances at G_0 for this same pair in our data, when the moments are calculated to the same white point. In other words the ratios of the luminances required for complementaries to form a white by admixture are the same as the ratios of the luminances at which they have the same brilliance as that white; $S_\lambda/S_\lambda{}^c = m_\lambda/m_\lambda{}^c$ but, as a function of wavelength, S_λ is not equal to m_λ. It is this relation of m_λ to S_λ that is missing from the CIE data and apparently not derivable from it. The fact of the identity of the ratios is remarkable in that, as in the case of the chromatic threshold, an external relation of identity of brilliance is found to be the same as an internal relation, in this case, that of cancellation strengths.

Since the ratio of the complementary moments can be deduced from our G_0 data, but the values for the individual wavelengths relative to each other cannot, the question of the manner in which these are distorted or lost in the derivation of the CIE system naturally arises. The answer to this question is not obvious. It seems most likely that the distortion occurs at the initial stage where the assumption is made that the units of the primaries can be treated as algebraic quantities. When a "match" is made for monochromatic light in the blue-green region (λ) by means of R, G, and B primaries, it is necessary to add amounts of the R primary to this wavelength until the mixture can be matched by some mixture of B and G, giving the equation $B+G=R+(\lambda)$ and this is then transformed to $B+G-R=(\lambda)$. In the original mixture, part of the cancellation (chromatic) strength of (λ) has been canceled by R, and then the total is expressed as though it were the sum of all; but the portion canceled is in terms of R and not in terms of the strength of (λ) by itself. From the nature of the G_0 function it would be assumed that they are, in fact, different. Thus the results would be distorted to the extent that the wavelength (λ) differed in strength from that of the primary R. This possible explanation is consistent with the fact that the major distortions appear to lie in this region and in the position of the minimum; it is also consistent with the fact that the system correctly predicts metamers since in the matching situation it is only necessary that the units be consistent.

An alternative approach would be to improve our data by making our ratios correspond exactly to those given by the CIE moments, but this gives no assurance of the absolute values for any wavelength. In the course of a search for some criterion that would permit an absolute adjustment we turned up a fact that, if confirmed, would offer a solution to the problem. As it stands it simply suggests an intriguing hypothesis. We found (Evans and Swenholt 1969) that for our data the sums of the complementary luminances at G_0 are roughly constant at a value very considerably *below* that of the white, roughly one-quarter of the luminance. The hypothesis would be that if the moments were so adjusted that their sums were constant for all pairs then each moment would be a constant multiple of the corresponding chromatic strength: $m_\lambda = aS_\lambda$. Our data neither justify the hypothesis nor suggest a value for a except for our conditions. It is perhaps sufficient to note that there may be a solution to the problem along these lines. To be useful, the solution must be shown to hold for at least all "white points," and hopefully for all points in the diagram. We have not been in a position to pursue the matter further. We can assume, however, that a standardized G_0 threshold function could be established by sufficient investigation and go on to consider the utility of such a function were it established. This raises the general question of the appropriate ultimate aim for a science of color perception.

We shall find in the next chapters that colorimetry is necessarily limited in scope to the specification of a range of possible perceptions, given all the stimuli involved. The limits of these ranges, however, are uniquely determined (for a standardized observer) by the physics of the stimuli. There is thus no reason to believe that all five perceptual variables cannot ultimately be predicted for these limits, given the stimulus and its environment. The question concerning the chromatic strength function can, accordingly, be restated by asking whether its availability would make this possible. In the present state of our knowledge the answer has to be that it apparently would not. Certified values for S_λ against a given white would permit us, as we have seen, to calculate the brilliance and saturation of any stimulus with respect to this white, and lightness and presumably brightness are already available, but we would not be able to transform to a different white nor to calculate hue. Even assuming that some simple conversion rule will be found for translation of the white point, the question of the requirements for hue prediction would remain.

The work of Hurvich and Jameson has shown that by using the Hering concept of unique hues and their chromatic valences it is possible to deduce what amounts to the chromatic strengths of the individual wavelengths, that is, our G_0 function. It is not possible, however, to reverse this process and deduce the valences from the chromatic strengths. We are in the same situation with respect to hue that we were in earlier with respect to the luminances of complementaries; the luminance ratios of complementaries can be deduced from the chromatic strength function but not vice versa, and the chromatic strength function can be deduced from the chromatic valences but not vice versa. We are left with the conclusion that what is missing from the CIE data is the Hering chromatic valence function or the equivalent. Since hue *can* be calculated from these and brilliance calculated from the derived chromatic strength function, it would appear that the ultimate solution must lie along these lines. I shall pursue this question further in Chapter 10, but it may be desirable to note that the above reference is to the *facts* Hering pointed out and which have been quantified by Hurvich and Jameson. It has nothing to do with the theories of vision that Hering and others have deduced from them.

We will find in the next chapter that both the chromatic strengths of the individual wavelengths and the wavelengths corresponding to the unique hues depend on the chromaticities to which the observer is adapted (the reference chromaticity). So we can conclude, in any case, that while the CIE system permits calculation of dominant wavelength, colorimetric purity, and relative or absolute luminance, it does not permit the added steps leading to brilliance and hue; some further psychophysical relation or relations must be added to the characteristics of the Standard Observer.

Eight

CHROMATIC AND DOUBLE SURROUNDS

The previous two chapters have been restricted to the appearances of stimuli seen against a surround which appeared achromatic. When the stimulus also appeared achromatic and had the same chromaticity as the surround, there were three perceptions distinguishable: the brightness of the configuration, the lightness of the stimulus relative to the surround, and the brilliance, evidenced by grayness for luminances lower than that of the surround. G_0 for this case was at a luminance match; $S = 1$. For chromatic stimuli the new perceptions hue and saturation appeared in addition to the others and G_0 as well as hue became dependent on dominant wavelength and purity. For all stimuli for which a hue could be seen we found G_0 greater than 0 and hence S greater than 1.0. All these studies were made with the surround at approximately 7000 K. We want now to consider the results obtained when the chromaticity of the surround was changed, and when both the central stimulus and surround were surrounded by a much larger stimulus, again at approximately 7000 K.

The first two experiments run after completion of the 10° surround instrument provided crucial information that set the pattern for much that followed; we can summarize the results here. The instrument had been designed to determine G_0 as a function of wavelength at photopic levels, with provision for use of the filters in the surround as well as in the main beam, both accurately controllable to calculated luminances. Intermediate purities were available in the main beam but not in the surround.

The filters for the instrument (mostly blocked interference filters) had been calibrated by high precision spectrophotometry; but to assure ourselves that no gross errors had been made, we placed each filter in turn in the *surround* beam, setting its luminance at 100 mL and comparing it with the central beam at 3000 K. This was done because it was known that a heterochromatic luminance match could be made more easily under these conditions. Both the ease with which the settings could be made and the appearances of the stimuli, however, surprised us so much that a summary was published as a Letter to the Editor (Evans 1967).

Briefly, we found that if any hue induced by the surround was totally disregarded and a setting made based solely on the point at which grayness disappeared, that point corresponded to the calculated luminance match point with the surround, independent of its wavelength and with good precision. Furthermore, the induced hues, although often quite intense, were usually easy to ignore because they fluctuated with eye movements (we tried to look mostly at the surround) and often appeared to lie in a different plane from the grayness, which appeared relatively steady. This experiment set the tone for much that followed, in that we decided to ignore hue and concentrate on grayness. It also posed a basic problem, since we knew that if the stimuli were reversed in position the zero gray point would *not* occur at a luminance match. The resolution of this problem came only many years later. We shall see presently that the surround, the stimulus which mainly controls the sensitivity of the eye, can always be considered to have a chromatic strength of unity ($S = 1$) for *any* chromaticity.

While hue was largely ignored in later work because we were not in a position to describe it accurately, it is worthwhile to note here the hue observations of this first experiment; they give rather strong support to the fundamental nature of the concept of unique hues. With respect to the surround (series were run at both 100 and 15 mL) the most noticeable fact was the relatively small number of hues seen once the adaptation had reached a steady state. While this was no doubt due in part to the absence of comparison fields, the impression was strong that only five or six hues were seen from the some 31 filters used. They were red, orange, yellow, green, and blue, and a color somewhere between blue and green. No violet was found from any of the short-wave filters after adaptation. No achromatic component was apparent from any of the filters, except possibly the yellow. The orange could be seen easily as a mixture of red and yellow and the blue-green, of course, as a mixture of blue and green. I experienced the others as distinctly single hues except, as noted earlier, I felt that the green tended toward yellow and the red toward blue. This latter tendency was so strong that I preferred to describe the hue as an intense pink rather than red, probably because of my long association of the word red with the long-wavelength end of the spectrum. This unique red (or pink) became a familiar sight to us in later years, often appearing unexpectedly as an induced color. In particular, in these experiments, the short-wavelength filters, while seen as blue, had outside the surround and within the central stimulus area an intense glow of this color. We came to the conclusion that the reason no surround appeared violet was that the eye had adapted to the red component, leaving only blue, but that this red component had spread diffusely to the outer and inner areas.

The appearance of the central stimulus was even more surprising to us; the induced hues from most of the surrounds appeared to split into two more or less independent components, and these components were themselves apparently always from adjacent unique hue pairs, blue and green, red and yellow, and so on. Occasionally orange appeared as a single hue. We received the strong impression that the hues induced into the central (always 3000 K) stimulus at times corresponded to almost independent responses eliciting the unique hues. It must be remembered that for light of such high purity in a 10° field, 100 mL is an unusually high intensity compared to that of most reports in the literature, particularly at the very short wavelengths.

This experiment also not only confirmed the spectrophotometry of our very dense filters (the lightest had a density of 2.3 and the darkest 3.9) but also indicated that my (RME) luminance settings corresponded to CIE calculated luminances with no apparent systematic departure in any region of the spectrum even under these conditions of rather intense chromatic adaptation. Since I was over 50 years of age at the time, this may suggest that in the short wavelengths at least there may have been some compensating factors.

In the second experiment we used the instrument as intended, with 3000 K in the surround at 100 mL, and determined G_0 for each of the filters in the center beam. We thus arrived at our first G_0 curve (1967), having had no previous idea of what it would be like. It was to be a long time before we learned its significance! Because it was obviously desirable that the results be at least comparable to those on which the CIE Standard Observer is based, filters were made up for the instrument that would convert the 3000 K to approximately Ill. C. With this filter in the surround beam, G_0 was then carefully redetermined for all the filters. These results were not published until the third article on the chromatic strength of colors (Evans and Swenholt 1969), in which we noted, "The effect is rather small; it consists mostly of a shift of the wavelength of the minimum in the yellow region, with little other change," and concluded, "This is an important result for color perception; it means that a color which appears grayed in daylight will appear to have about the same amount of gray when seen under artificial light. The result is confirmed by the small variation in appearance of color photographs . . ." We would now say that the shift in the wavelength of the minimum was *the* important difference between the two, a shift later found to be from around 588 nm for 3000 K to around 575 nm for 7000 K. In this later study 10,000 K gave 572 nm and 2000 K gave 592 nm. The minimum of the curve is thus dependent on the chromaticity of the surround, but all of these minima were seen as yellow.

It is apparent that a complete study of the effect of surround chromaticity on the G_0 threshold for all wavelengths and purities of the stimulus and surround would be an enormous undertaking. Furthermore we were not equipped to vary surround purity. Faced with these facts and the impossibility of devoting more than a few hours a week to observing, we tried to survey the field for major effects rather than to concentrate on any one phase of it. The results, while correct in their broad implications, are lacking in detail. The approach consisted of determining G_0 for a limited number of wavelengths (usually 17) at high purity in the stimulus, with a few high purity surrounds and then using the same set of filters in the surround for a few constant-wavelength stimuli. In addition, we made a number of studies of a single series of stimuli corresponding to a constant Chroma Munsell set. (See the next chapter.)

One of the major items of interest in all of this work was the brightness of the surround itself which was, in all cases, set at a calculated luminance of 100 mL. This brightness varied over a tremendous range, from a quite comfortable level in the vicinity of yellow, around 570 to 580 nm, up to a level hard to tolerate for the short wavelengths around 430 to 450 nm. These brightnesses paralleled the S values we had obtained against the white surround and could be predicted from them. If we assume that for an isolated stimulus the inherent chromatic strength is seen as brightness rather than brilliance, then yellow with its S value of 2 to 4 will be least bright; red and green at $S = 10$ to 20 will correspond to, say, 20 times the 100 mL brightness of white, or 2000 mL (a bright day); and 435 nm with an $S = 100$ or more will correspond to 10,000 mL or more, which is effectively direct sunlight on a white card on a clear day. This is a good description of the way the stimuli appeared. Since neither grayness nor fluorence could be produced in the surround by changing the luminance of any of them, the perception is in fact brightness, and not brilliance; and we have to conclude that, for an *isolated* stimulus, brightness depends on both luminance and chromatic strength. We have to qualify our description, however, by limiting it to a relatively small (in our case 10°) field against a dark surround. We shall have to consider later how this fact enters into the determination of the luminous efficiency function.

It is not only the brightness of the surround that makes the settings of G_0 difficult with highly chromatic surrounds; many startling hue effects occur, as already noted, especially with blue surrounds. There are, of course, afterimages of the stimulus which are mostly very strong, but these can be minimized by concentrating on the surround limiting the time of looking directly at the stimulus itself. Furthermore the grayness-fluorence threshold seems more or less independent of these phenomena and can be set with fair confidence after some practice. Each set of conditions, however, tends to be a new experience and any series becomes very fatiguing work. Many more replications than were possible for us would be required to produce precise data. We are convinced, nonetheless, that the data obtained are representative of the main facts involved as far as brilliance is concerned. We are also convinced that a systematic investigation of the hue phenomena involved, especially by some sort of inter-eye comparison technique, would establish many new facts that any theory of color vision would have to explain. A few rough observations will be mentioned later, but we might note here again, as an example, that the minima for each color temperature surround appeared yellow.

Representative regular G_0 curves for three surround wavelengths for observer RME are shown in Figure 8-1. Each has its own peculiarities and is quite different from the others and from the "normal" curve obtained against white. Several important and characteristic points are illustrated by these curves. The 528 nm surround shows no secondary minimum around 500 nm. This minimum is small but very well established against white and is strongly accentuated by the 608 nm surround. Furthermore the major minimum given by yellow against white does not appear at all with the 608 nm surround but does appear as a secondary minimum against 475 nm. None of the stimulus wavelengths shows a significantly higher G_0 than it gives against white except in the region around 575 nm; and

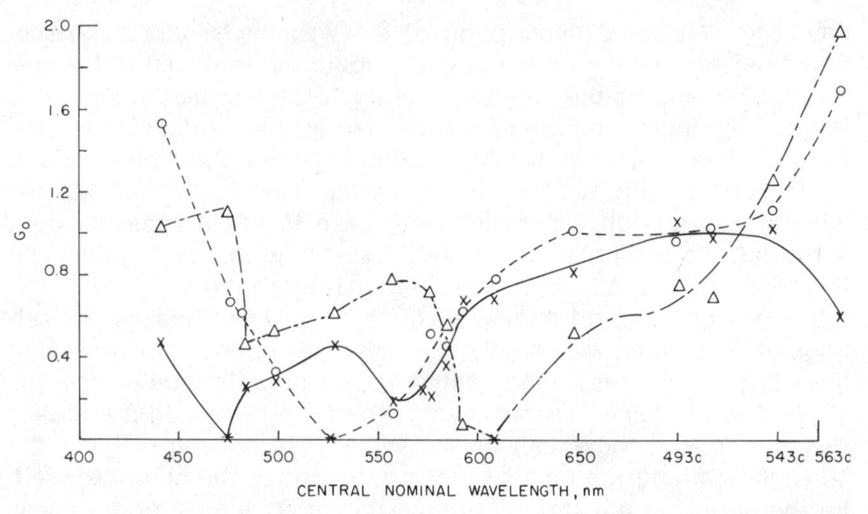

Fig. 8-1. G_0 as function of nominal wavelength for three high-purity surrounds: (×-×-×), 475 nm; (○-○-○), 528 nm; (△-△-△), 608 nm. (From Evans and Swenholt, *J. Opt. Soc.* **59,** 628–634, 1969.)

it is, apparently, a valid generalization that the maximum brilliance relative to the surround is produced by high purity stimuli against white except in the vicinity of yellow, but that this value is often reached when there is a large wavelength difference between the two. There is, however, no obvious relation between the chromatic strength of a wavelength when used in the surround and its strength when seen against white.

The most striking fact is, of course, that the major minimum for each curve lies at the wavelength of the surround. That each curve must have a zero point at the wavelength of the surround follows from the simple fact of metamerism. If the dominant wavelength and purity of stimulus and surround are the same, they are metameric at a luminance match ($G_0 = 0$). It is an oversimplification, then, to say that monochromatic yellow, or rather the wavelengths around 575 nm, have inherently lower strength than the others; this is true only for certain surrounds, the one most frequently seen being achromatic. It is apparent, then, that chromatic strength is not an inherent characteristic that can be assigned to the individual wavelengths but is an attribute of the relationship of a stimulus to its environment which can be described by a wavelength function for each situation. This suggests to me that the concept of adaptation as a simple adjustment in sensitivity of a number of receptors is too naive. It suggests also that a sort of field is set up by the adapting stimulus and that brilliance, at least, is somehow perceived as a sort of vector difference of the stimulus from the field; and this probably applies also to at least part of the hue perception. The reason that the field is relatively static for a given situation is, of course, that adaptation is an equilibrium condition between the tendency to decrease sensitivity when looking at the stimulus and to increase in preparation for the next (as in blinking) and, in our case, the influence of the dark surround outside the border. I am inclined to think that only hue studies can give a clear picture of the way in which differences between the response to the stimulus and this static "field" are evaluated perceptually. This is mere speculation but it does suggest that adaptation as a phenomenon affecting brilliance probably involves centers higher than just the retina.

More to the point is the fact that these studies give the first clear indication that brilliance as a perceptual variable is independent of lightness. This is demonstrated directly by a further experiment illustrated by the curves of Figure 8-2. In this experiment G_0 threshold determinations were made (by observer BS) for single stimuli for 17 different surrounds, all at 100 mL. It is seen that the value of G_0 can be made to vary from zero up to a high value characteristic of the stimulus simply by changing only the chromaticity of the surround. At each condition represented by any point on any of these curves the stimulus is seen as equal in brilliance to the surround, yet its lightness is determined by its luminance with respect to the sur-

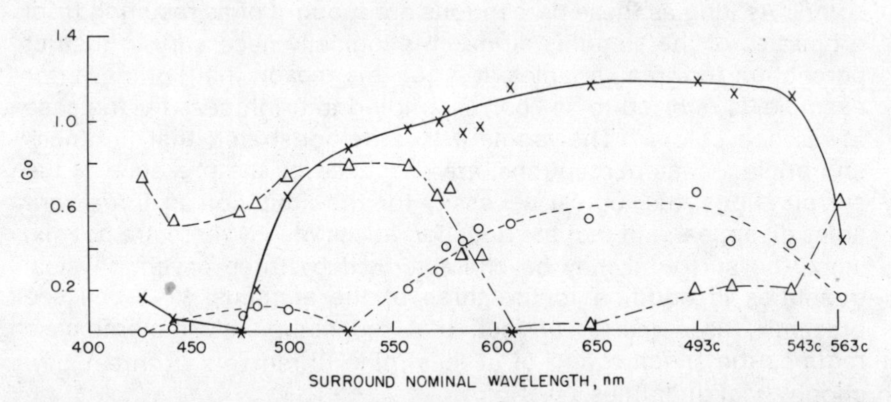

Fig. 8-2. G_0 as function of surround nominal wavelength for three high-purity central stimuli: (×-×-×), 475 nm; (○-○-○) 528 nm; (△-△-△), 608 nm. (From Evans and Swenholt, *J. Opt. Soc.* **59,** 628–634, 1969.)

round. Thus, the 475 nm stimulus appears quite dark (1/10 the luminance) against all surrounds with wavelengths longer than about 575 nm and through the whole extraspectral region. The lightnesses are comparable to the surround only in the neighborhood of the surround wavelength, and this applies to all stimuli.

It is necessary at this point to consider the general problem raised by the independence of lightness and brilliance. Ever since the fact has been known (long before Young) that all colors can be matched by mixtures of three others, and particularly since the establishment of three necessary and sufficient psychophysical variables for such matches, it has been a universal assumption that, *therefore,* there must be only three perceptual variables of color. This was stated explicitly by Helmholtz but has also, as far as I know, been assumed by all other workers, regardless of their other views. As a matter of fact, the three-dimensional requirement for perception is logical *only if* it is *also* assumed that the appearance of the stimulus is controlled entirely by the stimulus itself. And this is true only when the stimulus is itself the only thing affecting the eye, that is, the isolated stimulus case. The difficulty is that this *appears* to be true in *all* cases. The grayness, particularly of an object or surface color, but also for other cases, *appears* to be produced by the stimulus itself. By isolating the stimulus, it is easily demonstrated that this is not so, but the fact remains that it looks like a property of the stimulus. To a somewhat lesser extent this is also true of lightness. The darkness of a color is thought of as a property of the color; physically it usually is. However, *as a stimulus* it is simply dark *relative to other colors.* As long as these perceptions are thought of as resulting from properties of the stimulus alone, it is logically necessary to restrict perception to three variables. It is for this reason that lightness, for example, is referred to as "corresponding to brightness for the case of surface colors." The moment it is demonstrated that lightness and brilliance, as perceptions, are produced by the presence of the surroundings, the logical necessity for the limitation to three variables disappears; in fact the question arises of why there are not six, since the surround may be characterized by three psychophysical quantities in addition to the three of the stimulus. We shall see presently, as a matter of fact, that the completely general case requires the specification of at least nine (three sets of three) psychophysical quantities.

Once it is recognized that the appearance of the stimulus itself is changed by the environment in which it is seen (and this has been universally recognized) then it becomes an *assumption* that these changes must occur in the same perceptual variables that are controlled by an isolated stimulus, that is, that no new perceptions are introduced. Only a persistent desire to keep the system three-dimensional (so it can be visualized?) can explain the circumlocutions that have been resorted to, to make it so appear. Lightness and brilliance are produced by the *presence* of a second stimulus and are in addition to, and not modifications of, the three perceptual variables produced by an isolated stimulus. It was Helmholtz's attempts to prove that the perceptions were all due to the stimulus alone that led him into the conjectures for which he has been so widely criticized. It is my belief, based on my earlier quotation from his second edition, in which he essentially recognized the duality of the brightness aspect of heterochromatic matches, that he would have changed his views in time.

The recognition of brilliance as an independent variable requires that we reassess much of the previous literature. It means that in a situation such as that of real objects uniformly illuminated by light seen as achromatic, while the perception of brightness can be assigned to the illumination, we are left with four independent perceptual variables to describe the appearances of the individual objects. It means that in the comparison of any *two* nonmatching stimuli seen simultaneously (or in quick succession), and particularly if they are adjacent, they may differ in four perceptual attributes and not three. In other words, in the presence of more than one stimulus we are always dealing with systems that are potentially five-dimensional as far as appearance is concerned, but the number of potential dimensions is reduced to four if we restrict ourselves to perceived differences, and to three only if the surround for a given stimulus is constant and controls the sensitivity state of the eye. This latter case makes lightness and brilliance dependent on each other so that we can consider either as the independent variable. But a change in the chromaticity of the surround or a change in the chromatic sensitivity of the eye changes the relation of brilliance to lightness and so introduces a *new* three-dimensional system. The general case can thus be artificially reduced to four, but it cannot be reduced to three without danger of misunderstandings and a limited view of the whole. This is, essentially, the thesis of this whole book. We will consider the special case of object colors in the next chapter and more general cases in Section III. Here we want to pursue further the perceptions produced by two adjacent stimuli.

As noted earlier, all relative luminance points below a G_0 curve represent a luminance-wavelength stimulus that is fluorent against the surround for which the curve was obtained. From our studies on purity of the stimulus and the fact that a stimulus normally seen as achromatic will have a $G_0 = 0$, we know that all these points represent G_0 values for stimuli that are of the same or lower purity than that of the surround. We can, then, make the rather surprising statement that any stimulus with a dominant wavelength different from the surround and at any purity greater than zero will be fluorent at the luminance of the surround, the fluorence disappearing only at $p_c = 0$. This means that any stimulus placed on a chromatic background and having a lightness comparable to that background will tend to appear more brilliant than the background, the brilliance increasing with increasing hue difference. This hue difference itself will of course tend to be amplified by the situation. Colors of similar hue, however, will tend to be seen as grayish if their lightnesses are only slightly less than the background. Brilliance relations thus tend to amplify both the chromatic difference of spectrally nonadjacent hues and the brilliance difference of adjacent hues, in comparison with the background. These effects are seen as *added* to the lightness differences.

The very pronounced minimum in every G_0 curve and the fact that it coincides, for high purity colors, with the wavelength of the surround has the curious consequence that we can, if we like, assign an effective wavelength to a surround seen as achromatic. The true meaning, however, must be that the minimum occurs at the wavelength at which monochromatic light has most nearly the same brilliance as that of the surround and that the region of the spectrum from 570 to 590 nm is, chromatically, inherently weaker than other regions with respect to white, that is, is most *like* white. Against chromatic surrounds, however, this region can show considerably greater strength than hues nearer that of the surround. It does not in these cases act as though it had a considerable white content, so to speak, because the normally achromatic stimuli give $G_0 = 0$ for *all* surrounds. Expressed differently, the chromatic surround curves do not support the assumption that these wavelengths excite any considerable amount of achromatic response; they simply appear to have lower chromatic strengths.

A curious fact, incidentally, and one of no significance that I can see, was encountered in our study of surrounds of different color temperatures from 2000 to 10,000 K. If on the CIE chromaticity diagram the wavelength on the spectrum locus corresponding to our G_0 minimum is connected to the point on the blackbody locus corresponding to the color temperature of the surround, these lines are all approximately *tangent* to the blackbody locus! In other words the monochromatic wavelength having most nearly the same brilliance as a surround of a given color temperature can be estimated by extending the tangent from that point of the color temperature locus to the spectrum locus. For a color temperature around 5000 K this is also the shortest distance but it is not for other temperatures except those below 1000 K where the two loci coincide. It is regrettable that our inability to vary the wavelength of the stimulus continuously made determination of the exact wavelength of these G_0 minima impossible.

DOUBLE SURROUNDS

In work with a single surround, whether chromatic or achromatic, there are a number of limitations that make it a rather special case and prevent extrapolation of the results to perception in everyday life. There are two basic limitations and both arise from the fact that the stimulus that controls the eye sensitivity is seen in contact with the stimulus being studied. The first is the fact that the surround is necessarily seen as an isolated stimulus and hence can show neither grayness nor fluorence, that is, always has $S = 1$. The second is the fact of contact itself. In everyday viewing the stimuli adjacent to the one under consideration may have any relative brilliance, and the level to which the eye is adjusted may be caused by stimuli not in the immediate neighborhood. Specifically, for the achromatic case, the stimulus may be seen against white, gray, or black, and only the white is represented by the single surround. The more general case can be studied, and inferences for the general case drawn, by controlling eye sensitivity with a larger area outside of a single surround. In this way the surround itself may be made gray or black or any color at any level.

To learn the nature of the effects so introduced we made a considerable number of exploratory studies with such a setup. For these we surrounded the eyepiece of our instrument with a small circular fluorescent tube of approximately the 7000 K of our standard surround and covered this with translucent material to produce a uniform field somewhat larger than 45° and with an aperture just large enough so that most of our surround could be seen through it. We then filtered this to produce a good match with our surround at 100 mL. We were restricted to a single sensitivity level but could now produce in the inner surround all the perceptions we had formerly seen only in the central stimulus. This inner surround could be made to appear any shade of gray or a good black (equivalent to about 2% reflectance), or chromatic surrounds could be made fluorent, or could be set at G_0, and so on. The number of possible combinations is enormous. This work came near the termination of our investigations, and most of the data obtained tend to be sketchy and imprecise, but we felt that several facts were quite clearly established, at least qualitatively.

We first studied the effect of black in the surround. Black is such a positive and striking perception that we perhaps anticipated more effect from it than we actually found. All of these results can be summarized by the statement that the presence of the black between the outer and the inner stimulus simply weakened the effect compared with what we would have found had they been in contact. In other words the black simply acted as the zero stimulus it was, separating the two, and the fact that it looked black appeared incidental. When light of the same chromaticity as the outer area was introduced in the surround and its luminance gradually increased, this weakening effect started disappearing as soon as light was visible in the surround; and when the luminance matched that of the outer stimulus, essentially the same G_0 relations were obtained as with the smaller surround alone. The actual effect of the zero inner surround was to lower the luminance at which the G_0 threshold occurred by a factor of about four; that is, G_0 was increased by around 0.4 to 0.6. This increase in G_0 was accompanied by a corresponding decrease in the chromatic threshold, the ratio of the G_0 luminance to that of the black point remaining constant for all wavelengths and the same as for the single achromatic surround. Expressing it somewhat differently, we can say that the effect of separating the controlling stimulus and the one under study is simply to increase all brilliances in the latter. Introducing an achromatic stimulus (gray) between the two, decreases this separation effect and eliminates it at the match point.

The effect of placing stimuli on a black surround is often very striking, particularly with respect to the "brightening" of all colors and the increased visibility of details in the darker ones. These are often referred to as increases in "saturation." We see here that the effect is an increase in brilliance and an accompanying decrease in the threshold which permits lower luminances to be seen. In other words, it moves the perceptions toward those that would be seen from a stimulus lighted to a higher intensity. The question of whether there was an accompanying change in saturation is an important one which was not determined by our experiments since we worked with monochromatic stimuli.

If saturation is the perceived relation between L_w and L_λ in the expression $(L_w - L_{wt}) + S_\lambda(L_\lambda - L_{\lambda t})$, then if the effect of a black surround simply increases the effective total luminance there should be no change in saturation. If, however, the change is primarily in S_λ there should be saturation changes. With monochromatic stimuli and $L_w = 0$ there would be no change in either case. Experiments along these lines might settle the question of achromatic perceptions from monochromatic stimuli but if they did occur in our studies we were not aware of them because of our preoccupation with the brilliance threshold. They would have been neglected along with the hue changes both for this achromatic case and the chromatic one which follows.

The visual effect of high purity inner surrounds is necessarily complex because in this area brilliance is determined by the outer surround, and this in turn interacts with the central stimulus. Our investigations were restricted to monochromatic stimuli and primarily to the two obvious special cases, that of a luminance match of inner to outer surround and that of a brilliance match between the two (G_0).

In the first case, when a calculated luminance match was set for both, the inner surround was of course fluorent for all wavelengths to a degree dependent on wavelength. In spite of this there was little difficulty in setting the zero gray threshold in the central stimulus. The results can be summarized by the statement that the G_0's for this stimulus, calculated with respect to the outer surround, were in all cases intermediate between the values obtained with an all-white surround and those with an all-chromatic surround. A possible exception occurred with yellow which seemed to have little effect. It may be legitimate to assume that when the inner surround appeared fluorent it tended to take over adaptation. The obviously crucial question as to whether it had more effect on the hue of the central stimulus than it did on its brilliance was not answered.

The second situation, that of a brilliance match for all three, is of considerably greater interest in that the results appeared to be quite clear-cut and rather surprising. Rather than trying to set G_0 visually for the inner surround, which would have compounded the rather large errors of single settings, we set the luminance of the inner surrounds at the previously-obtained values for the stimuli against white. At this setting, for all wavelengths studied, the G_0 setting for the central stimulus was *independent* of the wavelength used in the inner surround and the same as if both surrounds were white. Thus, *under these conditions,* and for effectively monochromatic stimuli, there is no interaction of the S_λ values of the chromatic stimuli; they both remain the same with respect to the outer surround. Again the findings are marred by our total neglect of hue, but it seems safe to assume that large shifts must have occurred and suggests an important study that should be undertaken (necessarily by inter-eye comparisons) on the effect on hue of surrounds under conditions in which the chromatic strengths *do not affect each other.*

In a way, these findings constitute support, almost amounting to proof, for the general concept of brilliance as independent of the other perceptual variables. In general, when the three brilliances were matched, both lightnesses were less than that of the surround and sometimes but not always the same as the other, depending on the two wavelengths. The results are also suggestive of the importance in a complex scene of the adjacency of stimuli as compared to the overall sensitivity level of the eye, particularly if there are large brilliance differences between stimuli. If there is no brilliance difference, on the other hand, the lack of effect suggests a possible principle of harmony in color combinations. It is doubly unfortunate that hue was not noted for this case since it seems not impossible that this situation might maximize hue shifts, or at least isolate them so that they become more important relative to brilliance.

HETEROCHROMATIC MATCHING

We can now consider explicitly the problem of comparing two high purity stimuli of different dominant wavelengths. While all of our studies dealt with stimuli surrounded by others, it seems apparent that two stimuli seen as halves of a divided field will differ from this case only in the matter of which half of the field, if either, dominates eye sensitivity and perhaps in the degree of the effects.

Consider the case of a split circular field with monochromatic light of widely different wavelengths in the two halves. From what we have seen earlier, it is impossible to match these two for both lightness and brilliance simultaneously. Assuming that lightness is the perception of relative luminance, which seems to be the case, then at a luminance match one will usually be more brilliant than the other. On the other hand if a brilliance match is produced, one will usually be darker than the other. What will actually be seen under the two conditions will depend on the eye sensitivity level produced by the pair; assuming that this stays between what would be caused by either alone, at a luminance match one will be fluorent and the other grayish, but neither will be darker, and at a brilliance match one will be lighter and the other darker, but neither will show any fluorence or grayness. The two conditions are thus perceptually different and recognizable to a trained observer. This thus accounts both for the inherent difficulty of heterochromatic "brightness" matches (the brightness is not really relevant to either case) and for the fact that trained observers can learn to make luminance matches. We now encounter, head-on, the question "What is luminance?" We can only proceed by inference to determine how agreement on the subject was reached, historically.

THE LUMINOUS EFFICIENCY FUNCTION

Suppose that we have two monochromatic stimuli of the same wavelength in the halves of a split field. When these exactly match, the energy is the same for both. But if the two are different wavelengths, say from the short- and long-wavelength ends of the spectrum, then the energies needed to make them look alike in *any* respect are very different for the two. If we call the aspect for which they can be made to most nearly match brightness, then it is apparent that the energy-wavelength characteristic of this aspect is of fundamental importance to the subject. There are, basically, three ways of proceeding to determine this relationship: direct comparison of different wavelengths, consecutive comparisons with minimum wavelength differences, and comparison of each against a fixed "white." The first and third are found to be ambiguous, and the second is tedious and not sufficiently reliable. Gradually, however, over the years, it is found that there is in fact a function that *can* be obtained by all three methods and hence is fundamental. Such a function becomes fairly well established and the expert few learn how to make such matches. Furthermore, it is found that, using this relationship between wavelengths, the result of mixing two of them can be predicted with respect to this characteristic; that is, it is found that a simple additivity law holds when these relative values are used for each wavelength: the sum of any mixture of wavelengths matches visually for this characteristic an equal calculated sum for any other mixture. Such settings, however, are for experts only and there is real agreement only that such a function exists. Subsequently it is found that under certain special conditions if two spectrally different stimuli are alternated and the speed of alternation and the relative intensities are set so that flicker just disappears, the energy relations will correspond to those more or less agreed on by the experts and the settings can be made by relatively naive observers. A function can now be agreed on by all *and it is found to work* in the calculation of metameric matches. It is adopted as an international standard by the illuminating engineering world. Also adopted is a value for the energy of the wavelength with maximum efficiency for producing this characteristic (555 nm), and all other wavelengths can be expressed as having a fractional efficiency with respect to this. Using the additivity principle, the spectral energy distribution of any stimulus, and the efficiency per wavelength, we arrive at the summation of the products that gives the relative brightness of any stimulus. The adoption of a standard unit then gives a value for this characteristic for any stimulus. *This is luminance.*

This fanciful condensation of over 100 years of the history of colorimetry is not far from the truth. It must not be interpreted to mean, however, that the luminous efficiency function as we know it today is arbitrary except perhaps in minor details. It is, necessarily, fundamentally correct and represents a real and surprisingly constant characteristic of human vision. The point that we have to question is not the concept of luminance but the more or less universal assumption that the perception of luminance is "brightness" in all situations. We have already seen that "brightness" really consists of the two independent perceptions, lightness and brilliance, and need not repeat the arguments here. What is needed amounts to a further assumption, which seems to be borne out by all the facts, to the effect that brightness, by our definition, is in fact the perception of luminance *when it refers to the stimulus that is controlling eye sensitivity,* and not under other conditions. We have seen that the stimulus controlling eye sensitivity can always be considered to have a chromatic strength of unity ($S = 1$). When S is equal to 1, then, the luminance determines the brightness perception. We must conclude that the luminous efficiency curve itself holds for brightness for this condition only. When brightness divides into brilliance and lightness, that is, when both are present, then lightness is the perception of *relative* luminance.

There is an apparent difficulty here in the fact that luminance is one of the variables that is used to calculate metamers and that metamers hold for all nonextreme viewing conditions. This is not a real difficulty; the luminance of the stimulus cannot be used alone to calculate the appearance of a stimulus in a complex field, but only to calculate a stimulus that will match another when they are in contact or at least on the same background. The match is for *effective* stimulus *identity;* they match for all five perceptions, if all are present and the viewing requirements are met; they do *not* match, in general, if the background of each is different.

The concept that chromatic strength is unity for any stimulus to which the observer is adapted offers an apparent explanation of (and gains support from) two rather anomolous results reported by W. D. Wright some years ago. In the first, which we noted earlier, he reported an attempt to determine the purity threshold by adding white light to a monochromatic stimulus rather than the ususl inverse approach. He found that the threshold for the perception of the white addition was independent of the wavelength with which he started. This is consistent with their all having the same chromatic strength as isolated colors and with the independent perception of the achromatic component. In the other, he reported studies of purity discrimination (see Wright 1947) and found that, essentially independent of wavelength, the threshold for a purity (p_c) change was the same near purities of 0 and 1 but doubled at a purity of 0.5. Assuming that chromatic strength is unity for each mixture and again that the achromatic and chromatic components are perceived separately (as we must under our definition of saturation), this is what would be expected. At $p_c = 0$ white is the dominant perception and at $p_c = 1$ it is the monochromatic component. At $p_c = 0.5$ each is half as strong and so requires twice the increment to be seen.

DISCRIMINATION STUDIES

It is perhaps unnecessary to elaborate on the distortions in the interpretation of experimental results that have come about through the assumption of only three perceptual variables. They are, however, particularly notable in the case of studies aimed at determining the differential sensitivity of the eye to the psychophysical variables. Accordingly it is worthwhile to consider several of these types of study. Analysis in detail is not possible without more precise knowledge of quantitative brilliance relations, but we can indicate some of the problems by considering each of the psychophysical variables.

Suppose we have a split field with the same monochromatic wavelength in each half, maintained at constant matching luminance but with wavelength continuously variable in each. If we determine, for any wavelength of one stimulus, the just-perceptible difference as the wavelength of the other is changed, we have determined, for that wavelength and direction, what may be properly called the wavelength discrimination of the eye for those conditions. The major perceptual variable that has changed is hue. This is not, however, the only perceptual difference that has been introduced; our results indicate that there will now tend to be a brilliance difference also. Unless it can be demonstrated that this difference is subliminal and has no effect, it is not proper to consider a function so generated a hue discrimination function.

A similar situation exists for purity and luminance. A just-discriminable change in purity alone between two stimuli introduces a change in brilliance as well as saturation. This may not be immediately apparent because I have considered brilliance largely as a function of luminance, but a glance at the equation

$$L_s = (L_w - L_{wT}) + S_\lambda \, (L_\lambda - L_{\lambda T}) + L_{wT} + L_{\lambda T}$$

shows that G_0 can be set equally easily by purity changes as by luminance. (See the next chapter.) A purity discrimination function, therefore, cannot be considered to apply to saturation alone. And, of course, luminance discrimination is not lightness or brilliance discrimination alone. Brilliance thus enters into all discrimination studies in a manner which cannot at present be evaluated. That it is not a negligible factor is illustrated by the well known fact just-perceptible differences are not additive and cannot be compared. I believe it is a correct statement that, say 10, successive just-perceptible steps in any psychophysical direction do not result in the same perceptual interval in any two parts of color space except by accident of choice. This follows directly from the concept of chromatic strength, the multiplying factor S_λ. This factor does not depend directly on any of the three variables, λ, p_c or L, but is introduced by the state of the eye produced by the configuration being viewed. In general, judging a large interval is necessarily different from judging a just-discriminable step. I do not know whether just-perceptible differences would be found additive if the eye sensitivity could be held rigidly constant in both situations and am not aware that it has ever been tried. Finally, it does not seem out of line to suggest that the apparently non-Euclidean nature of the chromaticity discrimination functions may be due to the assumption that only three psychophysical variables are involved. The inclusion of the very irregular chromatic strength variable might well clarify the situation by making it four-dimensional.

It follows also from these considerations that no small-difference formula, based only on a transformation of CIE coordinates and with no term introducing the chromatic strength function, can predict equality of small differences in different parts of color space. The failure of the von Kries transformation, based only on a manipulation of the coordinates of the stimulus alone, is a case in point. The adjustment of the relative weights of the variables does not introduce a term based on their interactions.

EYE ADAPTATION

I have referred rather frequently to the sensitivity state of the eye under a given set of conditions and have included in this concept the sensitivity to both luminance and chromaticity. In the literature this sensitivity state is usually called the "adaptation level." I have avoided the term because I feel that the literature tends to use it to include too much and so obscures the actual facts involved. Or, perhaps it is better to say that the term has implied what appears to be an oversimplified explanation of too many phenomena. Thus there are at least three separable phases of perception to which the concept is applied: the change in appearance of both members of a metameric pair with change in viewing conditions (the main meaning), the adjustment to a new average level of stimulation from one scene to another, and the interactions between stimuli under the various states so produced. The implication is that all of these can be explained by a simple, not necessarily linear, adjustment of the absolute sensitivity levels of discrete receptor mechanisms in response to the spatial distribution of the stimuli and the recent time-of-exposure sequence of the eye. The facts would seem to imply that there is something more than these involved, that it is, perhaps, more a matter of some different kind of interaction of the receptor responses with the state thus set up that determines the perceptions, particularly those of hue and brilliance. This is not to say, of course, that the receptor outputs do not *also* change.

We cannot at the present time give a satisfactory explanation for these effects for the simple reason that we do not have a sufficient knowledge of what the effects are. This seems like a strange statement for a subject over 200 years old and yet it is true. Almost all investigations into color perception have been limited by the investigator's firm belief that he was dealing with only three perceptual variables, and even when the experiments were not actually set up to prove some theory of color vision, any generalized conclusions reached have been stated as either favoring or opposing one. In no case to my knowledge, at least in recent years, has there been a purely phenomenological study that admitted the possibility of more than three perceptual color variables, although this has been suggested by analyses of adaptation studies, for example, that by MacAdam (1956). Hue and brilliance, variables which we have seen to be heavily dependent on chromatic strength, while often recognized and mentioned indirectly, have been assumed to somehow necessarily follow from other facts rather than to be important in their own right. This is less true of hue, which has always been explicitly recognized as a perception, but only Hering and those with similar views have given it prominence in their work. We have seen that it led him directly to the concept of color valence that we have shown to be essentially chromatic strength, but no one seems to have carried this approach on to even imply that hue represents an independent fourth psychophysical variable introduced by the surroundings, that is, by the state of the eye, but perceived as a characteristic of the stimulus itself.

It is because all the results in the literature have been described in terms of three variables that we do not have sufficient facts available to write a clear-cut statement of the phenomena included in the term "adaptation level." It is true that Hering and others laid great emphasis on grayness and hue and that physicists have published many observations involving hue and saturation, but we do not know how to combine the two because each group has expressed its results in terms of a different set of three variables that were assumed to be the same. We now see that grayness is an aspect of brilliance, and that saturation, as it has always been used (although not always defined), has been an inextricable mixture of brilliance and of saturation as we have defined it, that is, the perception of effective purity.

Stated yet another way, I have avoided the use of the word adaptation because, in the literature, it implies a theory of color vision that the present treatment of the subject assumes to be incorrect. In all treatments of adaptation of which I am aware, there is the implicit assumption that all the perceptions resulting from a given stimulus in a given situation are due *only* to the sensitivity state of the receptors directly involved. As soon as we are forced to admit that one of the perceptions of such a stimulus is caused by an *interaction* with the response due to other stimuli, then the assumption and the theory are no longer tenable. Brilliance is such a perception. It is due to the sensitivity state of the receptors relative to the effects of the surround and not to the sensitivity state itself. Hence no transformation equation for the state of the receptors alone, no matter on what it is based, can predict the brilliance that will be produced by any *single* wavelength-energy distribution, although it will, of course, predict its metamers. Only if chromatic strength is explicitly present in the equation and two or more stimuli are considered, can the result of "adaptation" be predicted.

Such equations can probably not be set up on the basis of existing knowledge. Investigations leading to such a formulation, however, could be made by inter-eye comparison techniques. It would involve the use of a concept that, although it is not new, has not, to the best of my knowledge, been explicitly stated. I shall call the concept "adaptive metamers."

ADAPTIVE METAMERS

Metamers are defined as stimuli that produce exactly the same color response if placed next to each other in any environment or that can be substituted for each other without change in the color. This concept can be extended to two stimuli that have the same color when each is seen on a *different* background, the match now being caused by the combination of effects of each stimulus and its surround. If we restrict consideration to the central stimulus of each pair, then each of these represents a family of stimulus metamers but the two families are different. Adaptive metamers are thus relations between families of stimulus metamers. Each of these families can be specified by the use of three psychophysical variables, but the two situations each require the specification of six—three for the stimulus and three for the surround. Any formula that would predict the colors in such situations must therefore contain not only six terms but probably also terms representing their interactions. It is these latter terms that would be the object of the investigations. We can proceed a little distance in this direction on the basis of facts we already know.

THE G_0 CLASS OF COLORS

Consider the entire class of all possible colors seen at G_0 with respect to their surround; I shall call it the "G_0 class." Certain statements may be made that are valid for this class as a whole. They are, by definition, all at the same brilliance as that of their surrounds and hence none appears either fluorent or grayish. It is, however, a multidimensional class of pairs of stimuli and surround, because any stimulus can be brought to G_0 by changing either the *chromaticity* of its surround *or* by changing the relative luminances of the two. From this total class we can select all those that also match for *brightness* of the central stimulus, and these will now match for *both* brightness and brilliance. The colors seen within this restricted group will now still require three variables to describe them; they may differ from each other in hue, saturation, and lightness. What we do *not* know at this point is the extent to which these are now independent variables, that is, the extent to which specifying two now determines the third and thus restricts the number of families of stimulus metamers that can be matched exactly. If we assume, however, that for all stimuli there are at least two such pairs, since this would seem to follow from the concept of adaptive metamers, then to this extent the class of colors all at the same brightness and with neither fluorence nor grayness would be three-dimensional. The central stimulus against any one background can of course have any energy distribution and so exhaust all possible chromaticities, but we could not assume that any *one* surround chromaticity, chromatic or achromatic, would show all possible colors within the G_0 class. However, since no chromatic stimulus can be brought to a luminance match with its surround and remain at G_0, there is only a limited range over which two different combinations can produce the same lightness in the stimuli, and hence the number of adaptive metamers within the G_0 class may be quite restricted for any given combination. It is these ranges that need investigation because it is only within them that perceived color is five-dimensional; that is, it is only to the extent that hue, saturation, and lightness may be varied independently and still stay within the G_0 class that there are, with brilliance and brightness, five independently variable perceptions.

LIGHTNESS

Note that in the above discussion the crucial question is the range over which the lightnesses can be matched in such pairs. We need, therefore, to consider this variable somewhat more closely. As a perception, lightness has a different status than brilliance and brightness. While it is clearly seen that lightness is a characterisitc of the perceptions produced by a related stimulus, it is always apparent to the observer that this does involve at least two stimuli. In a sense it is more like a conscious decision than a direct perception; it rarely has the immediacy or the quality of forcing our attention that fluorence and brightness do. On the other hand, it cannot be dismissed as a perception because it is on the basis of matched lightnesses that heterochromatic luminance matches can be made, and there are many cases, such as when we speak of a dark color, that lightness is the important variable. It is in this sense, for example, that we speak of blue as being a darker color than red; it is always darker for the same brilliance. Furthermore, a group of stimuli all at the same lightness (matched reflectances) do show a similarity which may be important in the application of color.

Nevertheless, it is probably true that in most situations lightness is both the least important of the five variables and the one least directly associated with the stimulus as such. It is also so often associated with grayness that it tends to be obscured as a separate variable. It is, in fact, just this obscurity that has made it possible for us to accept for so many years the belief that the world of surface-color perceptions is three-dimensional; that (in our terms) lightness and brilliance are not separate variables.

ISOLATED STIMULI

Some light can be thrown on this subject if we consider the relationship between the manifold of perceptions produced by isolated stimuli and by the colors producible by the restricted class of G_0 colors seen against an achromatic surround. Each of these groups exhausts all possible physical stimuli, and yet they cannot be paired with each other perceptually because the related stimulus produces the perception of lightness and the isolated one does not. It is worthwhile to consider this more closely. Suppose we have a chromatic stimulus at G_0 with respect to its surround. The *surround* can be matched exactly by an isolated stimulus (necessarily by inter-eye or memory techniques) because as long as it is at higher luminance we have seen that it may be considered isolated. Thus it produces only the perceptions of hue, saturation, and brightness and can be placed in one-to-one correspondence with the other isolated stimulus. The central chromatic stimulus (at G_0) however, is always darker than the surround, a characteristic the isolated stimulus *cannot* have.

If we disregard this difference, which is not difficult to do, and consider *only its absolute brightness,* then it can be matched for *brightness* by the isolated stimulus, just as was done for the surround. This is the famous (or infamous) connection between brightness and lightness. The error comes in first neglecting lightness in order to make the match and then claiming that because the match could be made, brightness and lightness are the same variable, *or,* perhaps even more subtly, that because both the surround and the stimulus can be matched for brightness, lightness is just a relationship of brightnesses, and so there is only one variable. This latter is false reasoning. If the relationship is perceived *as such,* it is a new perception, not the same one. Brilliance is also such a perception but more obviously has to be accepted as a separate perception because of the obtrusiveness of the grayness and fluorence as perceptions.

The isolated class of colors can not be put into exact correspondence with even the G_0 class of colors against a single background except by neglecting lightness completely. Matching them for brightness only avoids the issue and, incidentally, is unconvincing for dark colors and impossible for black.

We have seen that the lightness of G_0 colors may be varied by a change in the chromaticity of the surround. Related colors that exactly match an isolated color for hue, saturation, and brightness can thus differ from it over a whole *range* of lightnesses. The class of G_0 colors at matched brightnesses is thus larger than the class of isolated colors by the extent of this range; it is four-dimensional rather than three-dimensional, with lightness the fourth variable.

THE CONSTANT-GRAYNESS CLASSES OF COLORS

We have been considering under the two previous headings the class of colors whose brilliances match those of their surrounds and hence show neither grayness nor fluorence. For every one of these perceived colors there is also a related series that match it for hue and saturation but which differ from it in brilliance either upward to show fluorence or downward to show increasing grayness until they eventually all become black. Each one of these levels corresponds to a different complete array of colors limited only by the applicable ranges of the other variables.

An early study in our experimental work (Evans and Swenholt 1968) throws some light on these classes of colors. In this study a thin strip of neutral filter was introduced into the surround of our instrument and appeared gray against the white. Any chromaticity in the central stimulus could then be varied in luminance until its color showed the same degree of grayness as the strip. Although the study was not extensive it was found that this degree of grayness occurred at a constant fraction of the luminance for G_0 for that chromaticity. This confirmed a similar result which had been found earlier with actual Munsell samples and which is discussed further in the next chapter.

We can assume, then, that each of these classes of colors is at a constant luminance ratio to the G_0 colors *on the same background.* Since this ratio also determines lightness, we see again that, for a constant background, lightness and brilliance are related perceptions but are independently variable by changing surround chromaticity.

It is apparent that the total class of possible colors from related and isolated stimuli combined is very much larger than that from isolated stimuli alone because for all chromatic stimuli there exists a whole class of colors for each discriminable brilliance level from black up to a luminance match. The total number in each brilliance class changes from a single response for all stimuli at black to a rapidly increasing number with increasing brilliance.

The question of the total number of perceptible colors is an interesting one which is somewhat outside our present scope. Judd, some years ago (Judd 1939), gave an estimate of 10 million. This was based on related colors using a fraction of the old National Bureau of Standards unit as the just-perceptible difference. Since it was based on experimental data it is likely that it is a satisfactory figure for stimuli seen against a single chromaticity. The extent to which varying the surround chromaticity would increase this figure depends on the above considerations and is, of course, unknown because of our lack of knowledge of the extent to which adaptive metamers are possible. Considering the approximations used to derive the figure, it may still be satisfactory, or it may be low by a factor of ten or more. It seems quite unlikely that it is too high.

Nine

VISUAL ARRANGEMENT OF COLOR STIMULI

Up to this point we have limited our study to relatively simple stimulus arrangements in order to examine the basic perceptions involved in perceived color and to get some feeling for their dependence on the psychophysical variables of the stimulus. We have distinguished five perceptual variables: hue, saturation, brightness, lightness, and brilliance. In order to account for these it has been found necessary to add to the usual luminous efficiency and color-mixture functions a further function that we have called chromatic strength. This has been found to be dependent on the relationship of the psychophysical variables of the stimulus to those of the surround, but the perceptions that it mediates are perceived as caused by the stimulus itself.

Because we know the details of this chromatic strength function only for special cases and then only its relation to brilliance, it is necessary to approach the whole subject from a different direction in order to attain a broader view of all the interrelationships involved. Such a different direction, or point of view, is offered by the large literature on the attempts at making orderly arrangements of stimuli according to the perceptions they produce, the so-called color-order systems. Outstanding among these, both for the talent and skill which have gone into the effort and for the results that have been achieved, is the Munsell system. For these reasons and because more colorimetric and perceptual data are available for this approach than for any other, I shall rely heavily on this literature for the considerations of this chapter. I do so with the intention of using it as a tool to clarify the subject and not with the intent of either criticizing it or suggesting an alternative method. It is perhaps unavoidable that both may appear to be implied. I shall refer to other published systems to illustrate certain points but will not be concerned with their relative merits.

THE MUNSELL SYSTEM

The Munsell system, like all others, is based on the assumption that three perceptual variables are necessary and sufficient to describe all possible colors, using the word "color" in its narrowest sense. We have found that three are not sufficient and so want to learn from the system and the studies related to it what difficulties have been encountered. Specifically we want to find out what it can add to our knowledge of the chromatic strength function as it is evidenced by brilliance and, perhaps, hue. Since the approach for the Munsell system, at least since soon after its inception, has been predominantly visual, all the perceptual variables must be involved actively in its construction and spacing. We want to analyze it to find out where they are and what roles they play.

The basic concept of the Munsell system, in line with its originator's desire to produce a "universal language of color," is quite simple, and the system has not been hedged in by restrictions on its use that might be thought necessary, other than the fairly recent requirement of a fixed-energy distribution for its illumination. This freedom has resulted in a fortunate dependence on appearance rather than theory, and in recent years its "improvement by committee" (of the Optical Society of America) has further strengthened this tendency. We can consider, therefore, that the present system is acceptable to a large number of experts in the field, within the limitations of the definitions involved in its arrangement. That many compromises have been involved is, I am sure, the consensus of all who have worked on it. Nevertheless the basic arrangement of the stimuli, except for hue, can be approximated by a quite simple psychophysical model, and I shall make use of this model for descriptive purposes.

The three Munsell visual variables are Hue, Value, and Chroma (capitalized to indicate that the words imply their definitions). Hue has the usual meaning but the Hue names that have been assigned are not in all cases in accord with general practice. This discrepancy has resulted from the somewhat too-scientific division of the complete hue gamut into 100 equally spaced hues and the consequent assignment of 5 equally spaced basic hue names. These are red (R), yellow (Y), green (G), blue (B), and purple (P). Intermediate between these are the pairs YR, GY, BG, PB, and RP, making 10 equally spaced Hue names. Each of the 10 divisions of the Hue circuit is then divided into 10 parts with the above 10 Hues occupying the center of each interval and numbered 5. The Hue 5 R is thus in the center of the R interval, with 2.5 R toward RP and 7.5 R toward YR. (These divisions correspond to the 40 Hues of the most complete Munsell Atlas so far published.) Note that this numbering is in the opposite order from that usually used for wavelengths. This arrangement of Hues and Hue names has the rather unfortunate result that the unitary hues are poorly represented, especially in simplified sets. On the other hand, it results in the best data we have on the dependence of hue on the psychophysical variables and provides us with a scale for wavelength against Hue.

Value is defined on the basis of the luminous reflectance of the samples as calculated, based on the CIE Standard Observer and Ill. C. The range from 0 to 100% reflectance is divided into 10 intervals based on equal visible steps and numbered 0 at black and 10 at white. All samples of the same reflectance have the same Value. A great deal of effort has gone into the exact spacing of this scale, based entirely on the neutral series from black through gray to white. The visual intervals of such a series depend heavily on the reflectance of the background on which the samples are viewed. Since it was desired that the spacing not be limited to a single background, the result is a systematic compromise and is defined by a quintic equation based on reflectance (see Judd 1952). The investigations on which it is based provide us with by far the best information available on the perception of grays against achromatic surrounds.

Chroma, the third variable, is defined as "the difference from a gray of the same lightness." All samples having the same assigned Chroma show the same total difference from a gray of the corresponding Value. All the samples can thus be thought of as divided into Chroma groups, each having constant luminous reflectance, each representing all hues, and each having the same visual difference from a gray of the same reflectance. The scale for Chroma is arbitrary, starting from 0 at gray and extending as far as realizable samples permit. Actual samples are given in two-Chroma step intervals and each of these intervals is intended to represent the same Chroma difference; that is, the Chroma scales have the same uniform spacing for all hues. It is apparent, I think, that this is the variable with which we will be most concerned.

These three variables are displayed in cylindrical coordinates with neutral samples as the axis, running from black at the bottom to white at the top, scaled in equal Value steps from 0 to 10. Planes perpendicular to this axis are thus constant-Value planes. Radial lines in these planes are loci of constant Hue. They are each scaled in equal Chroma steps from 0 at the axis to a maximum which is set arbitrarily by extrapolation of samples of the system to the calculated maximum visual efficiency nonfluorescent colors. Hue is spaced uniformly around the axis in equal 3.6° steps. Munsell color space is thus a closed system in cylindrical coordinates enclosing all theoretically possible nonfluorescent reflecting or transmitting colorants and based on these reflectances and transmittances. While the original arrangement, except for Value, was obtained experimentally, the arrangement has been smoothed and extrapolated, based on CIE coordinates of the actual samples. These values of the coordinates have been published as the Munsell Renotation system, thus permanently defining the samples. (See Judd and Wyszecki 1963 for a detailed description of the system.) It must be kept in mind, however, that, with the exception of Value, this operation has not affected the basic arrangement of the samples, which can still be considered as based on their perceived colors. We are thus free to use the system to deduce the dependence of the perceptual variables on the psychophysical ones, with the single constraint that Value indicates calculated luminous reflectance. We shall see that this latter is a fortunate circumstance. The original actual samples of the system have matte surfaces; the samples in the recent Atlas, based on Renotation values, have glossy surfaces to extend the gamut, which is limited to permanent pigments.

The cylindrical coordinate system permits separation of the variables into convenient two-dimensional arrangements which we shall use for description. Thus any constant-Value plane displays all Hues and available Chromas, any plane bounded by the axis and the maximum attainable Chromas is a constant-Hue plane, and any circle concentric with the axis in a constant-Value plane displays all Hues at constant Chroma and is called a Chroma circle.

In spite of its visual origin, or because of it, depending on the point of view, the Value-Chroma relations of actual colors can be approximated surprisingly well by a fairly simple, purely psychophysical model. Departures from the model are probably more significant than would at first appear, but the necessarily low precision of the placement of individual samples and their very limited range in terms of all possible chromaticities make deductions from these departures quite uncertain. In any constant-Value plane Chroma is approximately a constant multiple of colorimetric purity for any constant-Hue radius. The multiplying factor is a steep and irregular function of Hue and changes gradually at high Chroma for most Hues; it is also a linear function of Value, decreasing to 0 for all Hues at a point considerably below the 0 point on the axis. Thus in the whole space, all lines of constant chromaticity but varying luminance converge at the same point on the axis below the 0 point (about $1^1/_2$ Value units). In general, these lines are not straight in space but curve with both Value and Chroma. They curve, however, only with respect to Hue. If all the Hue planes are rotated about the axis so that they coincide, the loci of all lines of constant chromaticity are straight and intersect in a point. Value and Chroma are thus linearly related throughout to relative luminance and colorimetric purity with the latter an increasingly poorer approximation for the high Chromas. Hue, however, is not linearly related to dominant wavelength as might be expected; the locus of any constant dominant wavelength varies in Hue, both with Value and with Chroma. Thus, in general, all the samples in a constant-Hue plane have somewhat different dominant wavelengths, the differences changing progressively with Value and with Chroma. We shall concern ourselves primarily with only a few of these relations: the apparent linear dependence of Chroma on colorimetric purity, the linear dependence of Chroma on Value, and the curvature of the lines of constant chromaticity.

We noted in Chapter 7 that Nickerson and Granville (1940) had found that the colorimetric purities of a specially prepared set of 100 Hue Munsell samples, all at constant Chroma and Value, were approximate constant multiples of the purity thresholds for their dominant wavelengths as determined by other workers. Since this implied that they were also constant multiples of our G_0 thresholds, a number of studies were made to confirm this relationship. (See Evans and Swenholt 1968.) Briefly, we first confirmed that log $1/p_c$ for any Chroma circle is, in fact, approximately the same as our G_0 function; that is, the functions differed by an approximately constant amount from each other for any Chroma circle. This was based on the published Renotation data. We then calculated p_c settings for our filters that would correspond to a characteristic Munsell Chroma circle and determined their G_0 values. Although there were notable differences in the blue and yellow, we concluded from this work that, to a good first approximation, the requirement for constant Chroma as a function of wavelength was that all the chromaticities have the *same* G_0 value. Since G_0 is the log of the ratio of surround to sample luminance, this meant that the 0-gray colors themselves, corresponding to the chromaticities of such a circle, lie in a single Value plane and themselves define a G_0 Chroma circle. Since we had found that all constant chromaticity lines were straight (when projected on a Hue plane) and all intersected at a point, we could conclude that this G_0 Chroma circle defined the Chromas of a whole family of Chroma circles at all other Values. It was of interest, therefore, to determine the locus of all G_0 colors in the system since, within the validity of the approximation, this would define the whole system except for Hue.

By the use of published purities from a number of other Chroma circles, we determined experimentally the average G_0 values for the chromaticities of each. These average values were found to lie in a straight line in the Chroma-Value plane (projected), intersecting the neutral axis around a Value of 9.3 and decreasing in Value with increasing Chroma. The relations are shown in Figure 9–1. The approximate locus of all G_0 colors in the Munsell coordinates is thus a cone with its apex at 9.3 on the Value axis.

Note that in making this statement I am extending the Munsell system to include all possible related colors. Technically the system as published is restricted to theoretical reflecting (or transmitting) colors, but there is no reason for this limitation since placing a sample on a low-reflectance background can exceed these limits visually and fluorescent samples could physically. The coordinates are valid for all related stimuli.

The above linear relation between Chroma and Value for the G_0 colors had been predicted by the original study that had resulted in the discovery of the G_0 threshold (Evans 1959). In this study actual Munsell samples of Hue 5 R were studied, and it was found possible to select series of samples each member of which showed the same grayness against white as some member of the achromatic series. Each such series plotted as a straight line on the Value-Chroma (V-C) diagram (see Figure 9-1) and all were parallel. It was by extrapolation of these to higher Values that the existence of a G_0 series was deduced and later confirmed by the use of an aperture instrument.

By combining these results with those above and the other studies already reported, we find that the loci of the colors of constant grayness (gray content) are also cones with their apexes on the neutral Value axis, hence parallel to the G_0 cone. By extension, the colors of equal fluorence lie on similar cones above G_0, but we have no data on these and they must become deformed as the luminances

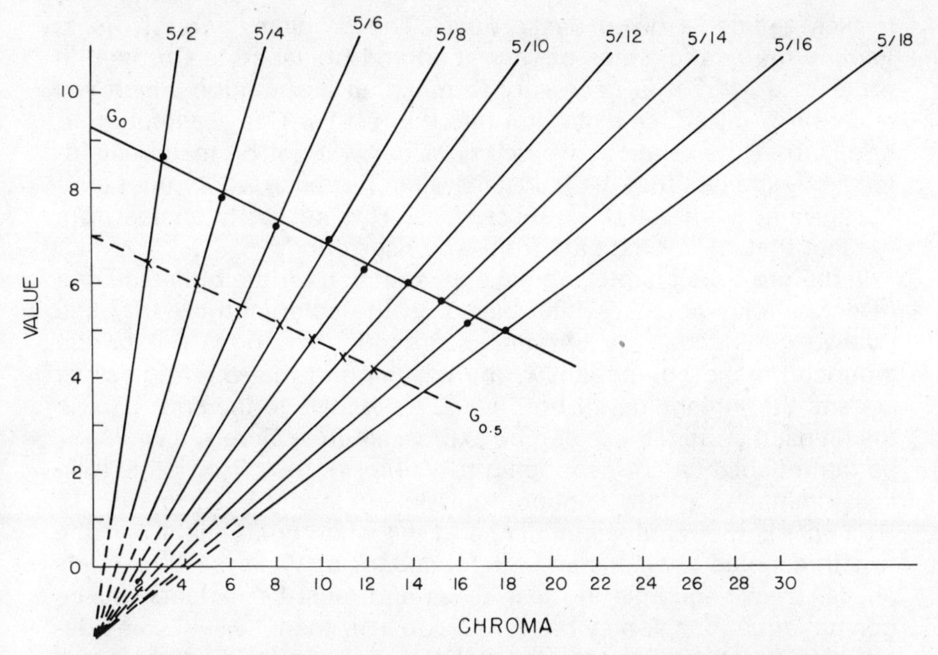

Fig. 9-1. Lines of constant chromaticity in the projected Chroma-Value plane, designated by their Chroma intercepts in the 5 Value plane. Solid G_0 line calculated from experimental data. All lines represent cones around the Value axis. Dotted $G_{0.5}$ line is for constant gray content. (From Evans and Swenholt, *J. Opt. Soc.* **58,** 583, 1968.)

exceed that of the surround. Note that this construction contributes nothing to the positioning of Hue in any Chroma circle. This varies smoothly with both Value and Chroma without regard to these relations. Expressed differently, these linear relations apply when all Hue planes are rotated about the Value axis until they are all coincident.

We want to consider the implications of these facts and, particularly, to use them to discover how the perceptions of lightness, saturation, and brilliance are represented in this system, since all must be present.

SATURATION, BRILLIANCE, AND CHROMA

We note first that by definition all the colors in a given Value plane have the same luminous reflectance and hence have the same lightnesses against a given background. The problem therefore is to deduce the relationships of saturation and brilliance to Chroma. In order to do this it is necessary to make an assumption which will necessarily introduce distortion into the results. This assumption is, briefly, that the relations we will consider will not be materially affected by the fact that the Munsell system has been worked out for a background with variable reflectances. This necessity comes from the fact that all our data are from white surrounds.

In the previous chapter we gave an equation for the brilliance of a given stimulus against white, based on the colorimetric purity, the relative luminance, the chromatic strength ($S_\lambda = \log^{-1} G_0$) of the monochromatic component λ, the assumption that $S = 1$ for white (w) and the thresholds for both λ and w, relative to the surround. In the Munsell system these can be expressed quite simply. If we let R_V be the reflectance corresponding to Value (V), then $R_V p_c$ is the relative luminance of the monochromatic component (surround = 1), $R_V(1 - p_c)$ is the relative luminance of the white component, and the two thresholds are constants, independent of *V*. These thresholds are far from negligible in most cases and must be included in the general equations. It may be well to consider them here in some detail. We noted earlier that the achromatic threshold in our work came at about log 1.6 below the surround, and this corresponds to a reflectance of 2.5%. This holds, therefore, for the white component of all samples. Since this is comparable to the surface reflectance of a real matte reflecting sample, we should note here that this surface reflectance is presumably included in the spectrophotometry of the sample itself and so does not enter explicitly; it simply acts as a limitation on attainable p_c. This question of the adequacy of spectrophotometric methods is, of course, an underlying problem in the study of all perceptual color in relation to the stimulus variables. We assume here that for *matte* samples the correspondence is good.

The threshold for the monochromatic component was found to be constant at one log unit below its G_0 value for all wavelengths. This threshold can thus be expressed as $^1/_{10}\ S_\lambda$ and again is constant for a given wavelength. This constant is important not only for the rather large effect it has on the system but also for the fact that it justifies our definition of saturation. S_λ is the ratio of the luminance of the surround to that of the stimulus at the G_0 threshold, and we saw earlier that this threshold for the monochromatic component did not change if white light was added. Thus when white is added, sufficient to bring the total luminance to that of the surround, $^1/_{10}$

S_λ is the colorimetric purity threshold for the stimulus. Since this threshold is determined *within* the stimulus and S_λ is determined by comparison with the *surround,* it provides satisfying proof that the white and the monochromatic components are, in fact, seen quite separately. The definition of saturation as the perceived relationship between the two thus becomes entirely reasonable.

We can now set up an equation for the brilliance (strictly, chromatic strength) of any sample in the Munsell system. If we let R_V be the reflectance corresponding to its Value, p_c the fraction of the reflectance that can be considered monochromatic (the dominant wavelength), S_λ the chromatic strength factor ($=\log^{-1} G_0$) for this wavelength against a white surround ($R_V = 1.0$), 0.025 the threshold for the white component, and $^1/_{10}S_\lambda$ the threshold for the monochromatic component, and assume that the brilliances are additive, then we can write:

$$\text{brillance} = R_V \cdot S_\lambda \cdot p_c - {}^1/_{10}S_\lambda + R_v(1 - p_c) - 0.025$$

MUNSELL CHROMA

What we want to learn is the perceptual concept corresponding to Munsell Chroma and its relation to brilliance. There are a number of possibilities, among them the sum of saturation and brilliance, and saturation alone. The most obvious assumption is that "the difference from a gray of the same lightness" simply means the brilliance difference. To test this assumption we can set up an equation for Chroma in these terms and see whether it fits the facts of the system sufficiently well. Such an equation simply assumes that Chroma is a simple multiple, a, of the brilliance difference:

$$C = a[(R_v \cdot S_\lambda \cdot p_c - {}^1/_{10}S_\lambda + R_v(1 - p\ ) - 0.025) - R_v]$$

As far as the Munsell data are concerned this equation contains two unknown constants, *a* and S_λ. There is also the problem that the data are in terms of constant Hue rather than dominant wavelength, but this can be minimized by choice of a color for which S_λ is changing slowly with wavelength. The value of S_λ for any Hue, as exemplified by the system, can be determined by taking advantage of the fact that at G_0 the brilliance of the sample is unity. By use of our data for the Go's of the different Chroma circles and the Renotation data for a number of them, it is found that S_λ is, in fact, quite constant in the data. By substitution of this value in the above equation for a whole series of Chromas for that Hue, it is possible to check the *constancy* of the assumed factor, *a*. These computations were carried out for a number of Hues and it was found that the factor *a* was in fact sufficiently constant to indicate that this concept of the perceptual nature of Chroma is indeed the correct one; that is, that constant Chroma is a constant brilliance difference from a gray of the same Value. No other formulation tried met this requirement at all well. The value of the constant *a* is approximately 30.

Although the implications of this equation for the Munsell system as a whole are somewhat outside of our present scope, it is interesting to consider them briefly. For instance, Chroma is not strictly linear with p_c the departure varying with the value of S_λ; the intercepts on the V axis are not at 0 and 10, as might be expected, but are affected by the threshold values. It is also of interest that the equation does not indicate that, in a given Chroma circle, $1/p_c$ is strictly proportional to S_λ as the original similarity of the log $1/p_c$ and Go curves implied. The log $1/p_c$ curves for different Chroma levels are not strictly parallel, and the system data do, in fact, show this effect.

SATURATION

Of more immediate concern to us is the light that these considerations throw on the relationship between saturation and Chroma and the nature of the saturation perception itself. We described saturation earlier as the perception (in the monochromatic plus white metameric appearance of a color) of the brilliance of the monochromatic component in relation to its total brilliance. The above considerations lead us to a rather new concept that can be called the "brilliance purity" (p_b) of a color. Saturation can then be defined as the perception of this variable. For the Munsell coordinates this becomes:

$$p_b = \frac{R_v \cdot p_c \cdot S_\lambda - {}^1\!/_{10} S_\lambda}{R_v \cdot p_c \cdot S_\lambda - {}^1\!/_{10} S_\lambda + R_v(1 - p_c) - 0.025}$$

In earlier chapters I referred to it as the "effective" purity.

This concept is in good agreement with general experience and explains the obvious fact that two colors of different dominant wavelengths can appear equally saturated at very different colorimetric purities. On the other hand it has some interesting consequences in terms of the literature on the subject. The value of S_λ varies from 1 for white through 2 to 4 for yellow up to well over 100 at short wavelengths. The ratio of monochromatic to white brillance in a sample can thus cover a tremendous range. Saturation discrimination, however, is relatively low, the generally accepted number of discriminable steps from a purity of 0 up to 1 being around 8 for yellow with a maximum little over 20 in the blue. The equation suggests that while for low values of S_λ the maximum saturation may be set by a purity of 1, at high values of S_λ the threshold for the perception of the white may be passed long before this purity is reached. This gains some support from the Munsell Renotation data itself. For the sample 10 PB 4/30, for example, with a dominant wavelength of 441.6 nm, the colorimetric purity of this extrapolated Chroma (at the MacAdam limit) is only 0.1187! We do not know what the saturation of a Chroma of 30 would look like but it could very well be beyond the maximum for that wavelength; that is, further Chroma changes would no longer show a saturation difference but only monochromatic brilliance. We thus conclude, as we did earlier, that there is a maximum for saturation, even on a white background. This conclusion conflicts with the many statements in the literature that saturation, even at a purity of 1, can be increased by a change in eye adaptation, which may be true for yellow but seems very unlikely for blue; we have no evidence one way or the other because the literature does not distinguish between brilliance and saturation. What *is* certain is that *brilliance* can be increased at any purity, including one, by such changes. Within the range in which saturation discrimination is possible this also should be dependent on the surround. Verification of this fact must await experiments in which *both* perceptions are taken into account in describing the findings.

It might be well to note here, again, that a close relationship between saturation and brilliance, even though they are clearly independent perceptions, is indicated by the equations. The independence is shown by the fact that S_λ is dependent on the surround, while p_c is not and they both enter both equations. Brilliance, however, is a sum and saturation a ratio. The independence is most apparent for $p_c = 1$, which puts no constraints on brilliance. For a *given* surround the relationship is close; for the general case it is not.

One of the major criticisms of the Munsell system in the past has been directed at the apparent increase in lightness as Chroma increases in a constant-Value plane. We have seen that Chroma depends primarily on brilliance but, by definition, lightness is constant in such a plane. As Chroma increases there is thus an increasing conflict between the brilliance and the lightness perceptions, and failure to recognize them as separate perceptions leads to confusion. At high purities, of course, it leads to the usual problem of heterochromatic matches. It was just this conflict and the consequent almost complete failure of agreement between observers of high purity samples that led to the arbitrary choice of constant calculated luminance as the *definition* of a Value. The whole situation indicates again, of course, that the problem is four-dimensional.

The Munsell system thus affords us a clear confirmation of the concepts we have been developing and at the same time gives us a good insight into both the quantitative and qualitative relations involved. Our main interest lies in the perceptions themselves rather than the system, but much can be learned by considering the general problem of color-order systems of which Munsell is one example. The general problem is to arrange all possible stimuli in such a way that the independent perceptions they produce are clearly displayed in an orderly arrangement. The perceptions can then be scaled in terms of the arrangement coordinates. For simplicity I shall limit the discussion to stimuli represented by reflecting samples, but this is a limitation on the *range* of possible stimuli *only;* it does not otherwise affect the principles involved.

All attempts at solving this problem have been based on the unquestioned assumption that a complete solution in three dimensions exists. We have seen that it does not. This is the reason why there have been "competing" systems and the reason for the strong preferences of individuals for one arrangement or another. All such arrangements are necessarily compromises in which one set of three perceptions is emphasized.

Although most of the systems that have been proposed rely so heavily on the stimulus variables that they are of little interest here, it is worthwhile to look briefly at the system published by the Container Corporation of America as the "Color Harmony Manual." While this is based wholly on theoretical considerations with regard to the stimulus (Ostwald–Hering), it leads to a somewhat different emphasis on the perceptual variables.

THE OSTWALD SYSTEM

The Ostwald system, as I shall call it, is also an arrangement in cylindrical coordinates with a neutral axis, hue constant in radial planes, and saturation increasing outwardly from the neutral axis. The characteristics of particular interest from our standpoint are the nature of the hue spacing and the emphasis on brilliance rather than lightness.

The hue spacing in this system is based on the Hering concept of the four unique hues and their bipartite mixtures. The hue circle is divided into 24 parts. The unique hues (six divisions apart) are thus all present, as are their equal-part mixtures (halfway between). The hue circle is thus more logical from the appearance standpoint than is the Munsell spacing based on hue discriminations, and it is somewhat easier to name the colors by their appearances.

Each constant-hue plane is bounded by an equilateral triangle with the neutral axis as one side. The apex opposite this side is the color with the maximum obtainable freedom from perceived "whiteness" (saturation) and "blackness" (brilliance). In practice, the samples are somewhat short of this goal, but it is apparent from our standpoint that if they were attainable these would be the MacAdam limit colors of this hue at G_0, and we can so consider them. The series from this "full color" in any triangle up to white is a series in increasing whiteness; that is, decreasing purity and increasing lightness. It is apparent that if the full color were at G_0, this whole series would be also; it would be a series of decreasing saturations at constant brilliance. From the "full color" apex to black, lightness, brilliance, and saturation all decrease together to form the "blackness" series. Actually this is accomplished by using essentially the same colorimetric purity series as that for the "whitenesses" so that the colors in any line parallel to the neutral axis all have approximately the same colorimetric purity. Since this simulates what occurs when the illumination on a fixed reflection sample is varied, it is called a "shadow series."

It would be out of place here to attempt to carry our analysis further but it is apparent that in this system lightness is a variable very much subordinate to brilliance and that hue is spaced more nearly according to its appearance than it is in other systems. The rather decided preference of many artists for this system lends emphasis to the importance of these variables in the use of color. On the other hand the Munsell system gives, in a way, a more complete isolation of the perceptual variables by combining saturation and brilliance into the single variable, Chroma. Both systems are necessarily compromises of the impossible problem of representing four variables independently in a three-dimensional space. Such an arrangement, by definition, requires four coordinates and hence a four-dimensional space. In the real world we are limited to a set of three-dimensional sets, and it is of considerable theoretical interest to consider different compromises that might be of value. The problem will serve as introduction to a fascinating compromise suggested and actively pursued by the late Dr. Judd as the "Ideal Color System."

POSSIBLE COLOR-ORDER SYSTEMS

All color-order systems, as the term is usually intended, deal with related colors. The isolated color case, in which the sample is isolated instrumentally and a visual match produced in the instrument is usually known as visual colorimetry and is of little practical interest. In the related color case the stimulus *per se* can be defined by three psychophysical variables and the same is true of the "surround." In the completely general case it is probably necessary to also define a third stimulus, again with three psychophysical variables, that determines the adaptation state of the eye to the extent that this can be considered constant. The general stimulus for related colors thus has a *minimum* of six independent variables. We have seen that this general stimulus produces five independent perceptions. Only one of these, brightness, is seen as characteristic of the whole. The other four are seen as characteristic of what I call the central, or just *the* stimulus. While the general brightness level undoubtedly affects all the other perceptions, particularly with respect to their ranges, we can neglect it here simply because the stimulus involved is the illumination of the real stimuli and so needs only to be specified, whatever the arrangement.

If we consider only the stimulus and its surround then, there are six psychophysical variables involved. These can be stated as the dominant wavelength, the purity, and the luminance of the stimulus, and the *relative* luminances, dominant wavelengths, and purities with respect to the surround. That this set of six produces only four perceptions is due, apparently, to the fact that hue and saturation of the central stimulus are simply *shifted* by the surround without producing a *new* perception, only lightness and brilliance being new perceptions *due to* the interaction of stimulus and surround. Since a generalized color-order system must consist of a *series* of three-dimensional sets of stimuli, our problem boils down to deciding which of the four variables will be held constant in each three-dimensional set, while varying between sets. We saw in the previous chapter that all possible perceived colors could be divided into three-dimensional families of colors, each family having constant brilliance. This is only one obvious possibility; among others is that hue could be held constant in each set.

When we come to consider practical realization of such sets even constant-brilliance sets (which are the only ones I shall consider) present real practical difficulties. The fact that brilliance is controlled not only by the luminance of the surround but also by its chromaticity makes lightness and brilliance independent variables. While we can probably say that the brilliance of the surround can always be taken as unity, we do not yet have any simplifying principle by which the infinity of possible surround chromaticities can be reduced to a manageable number. There undoubtedly is such a principle but it is not apparent from our work. The completely general solution is thus not possible to state at the present time. Nevertheless it is of interest to consider a few special cases, although even the simplest are too complex for more than a sketchy treatment.

We can start by noting the reasons for some of the complications. There is, first of all, the question of how the system is to be used. It was Munsell's concept that a collection of color stimuli could be assembled that would produce uniformly scaled perceptions of the color variables along three axes, that these scales could be memorized, and that any color perception could then be described by its position on these three axes. Alternatively, the perception could be looked up in the "atlas" and defined rigorously. If it had been true that only three (or four with brightness) variables were involved, it is safe to say that all problems would long since have been resolved. We have seen that it is not always possible to find such a matching perception even in a complete atlas as long as it is on a single background. In order to match any perception it is necessary to introduce a different surround either for a sample in the atlas or for the stimulus causing the perception. This causes two problems. If a different surround is placed about the atlas sample, a match may be produced but it is no longer legitimate to describe the perception by its position on the atlas scales. At best the combination describes a *method for producing* the perception. If, on the other hand, the stimulus producing the perception is surrounded by the same background as the atlas, a match can, of course, be found but now the original perception has been changed. For some purposes this is legitimate since the system is now being used to locate a Munsell *metamer for the stimulus,* but this abandons the whole concept of describing the *perception* of the stimulus. It is at least theoretically possible for an atlas to cover the entire gamut of attainable reflecting stimuli, but if it is used for this purpose, perceptual scaling becomes irrelevant, and psychophysical scales or those based on mixture primaries might as well be used. In effect, it reduces the system to a form of a visual colorimeter.

A second complication arises from the physical limitations of reflecting stimuli. These limitations are of several kinds. If the atlas samples are all at one illumination level, then theoretically it contains matching samples for any reflecting stimulus at the same level but not at others. This is the problem of the artist attempting to copy a natural scene. If, for example, the reflecting sample in the scene is such that multiple reflections with the same surface can occur (gold-lined goblets, some fabrics, etc.), then the purity of the stimulus can exceed that of a singly reflecting sample. If the atlas samples are matte surfaced, attainable purities will be limited by the outer surface reflection, and this limitation may not apply to the stimulus (glossy surface, transmitted light, etc.). Such limitations, however, simply mean that the *range* covered by the atlas is inadequate; presumably the stimuli could still be described by extrapolation of the scales.

In spite of these complications and limitations it is not necessary to abandon Munsell's basic concept of describing a color perception by its position on scales (there are now four rather than three such scales), each representing a single perceptual variable. The problem is to produce examples to which reference can be made.

One interesting possibility is suggested by the earlier considerations. We saw in the previous chapter that it is possible to collect all perceived colors into unique three-dimensional sets in which all the stimuli of each are at constant brilliance (grayness, fluorence, or neither) with respect to their surrounds. We saw also that, at least for achromatic surrounds, changing the luminance of the surround changes this brilliance by the same amount for all stimuli regardless of their chromaticities. Furthermore, the "black point" against white was found at a constant luminance ratio below the surround luminance, for such stimuli. These facts imply that if the stimuli could be produced, a set of colors, all of which appeared black against a white surround, could be made to represent a whole series of constant-brillance sets simply by surrounding the samples with different reflectance achromatic backgrounds. Such a set is obviously impossible in reflection stimuli, but this possibility of varying the brilliance of the whole set by background reflectance would be characteristic of any set at constant brillance and there may be some utility in the concept. At least it would enable us to find sample conditions that evoked three of the required four variables along independent scales; that is, hue, saturation, and brilliance. Such a stimulus set would, of course, be two-dimensional; the only reason to have more than one would be to overcome physical limitations. The arrangement in the set would logically be circular (polar coordinates) with hue constant along radii and saturations spaced uniformly from the center. Lightness thus becomes completely dependent on hue and saturation, decreasing with increasing saturation and varying widely with hue in the same manner as the G_0 colors. In such a set, unfortunately, hue would change somewhat with brilliance.

As we saw earlier, lightness in this situation can be varied independently by changing the chromaticity of the surround. Such a change in the surround color, however, now upsets everything. We cannot, in fact, even state what would be likely to happen because we do not know how to predict the new hue, saturation, and lightness relations. It is possible that the relations are fairly simple but they would have to be established by new investigations.

This raises again the question of the importance of lightness in perception. In a set such as those above, against any achromatic surround, there is no question that high saturation colors would look darker than the low saturation ones, in some hues very much darker, and that an external stimulus might have a different lightness against its background while still matching for the other three perceptions. How important is this if it is recognized that the difference can exist? Would a simple estimate of its lightness (apparent reflectance), made perhaps with the aid of a neutral scale against white, be sufficient? If so, then the whole approach reduces to a single three-dimensional array in which the same sample chromaticities and reflectances are given against each of a series of achromatic surrounds; that is, the surround rather than the samples represents the third variable. Only experience would tell whether this would be superior to the Munsell arrangement in which brilliance is the dependent variable.

The system, if I may call it that, would have one marked advantage over any yet proposed in that it would provide a logical descriptive position for the fluorent colors; no other system does. Although our work does not prove it, presumably all colors of constant fluorence would be produced by the same set that produces constant grayness. Note, however, that in order to produce the required purity—lightness relations for many hues in actual reflectance samples, physical fluorescence would have to be present and that this in turn would place new requirements on the energy distribution for the illumination of the "atlas."

JUDD'S "IDEAL COLOR SYSTEM"

The possibility suggested by Judd (1969) deals with a different phase of the subject, that of the actual scaling of the perceptual variables themselves. Although our purpose is to determine the nature and interrelations of the variables to be scaled, the difficulties encountered in attempts at scaling demonstrate directly the inadequacy of the assumption of only three perceptual variables and are thus of direct interest to our subject.

In the article cited above, Judd reviews in some detail the problem of arranging color stimuli in a three-dimensional space in such a manner that the colorimetric distance between any two stimuli will be proportional to the perceived color difference between them. He defines such an arrangement as Ideal Color Space and considers the reasons why it is generally accepted that there is no Euclidean solution to the problem when the stimuli are to be seen against a single surround. He then considers the somewhat less ambitious problem of having all *equal* distances represent the *same* visual difference and reaches the same conclusion. Finally, he considers the effect of varying the background against which the stimuli are seen, as a function of the stimuli themselves, pointing out that, contrary to popular belief, this is one of the principles on which the Munsell system is constructed.

The Value scale spacing of the neutral stimuli of the Munsell system was obtained not by determining equal visual intervals on a single surround but equal intervals when the surround was halfway between the reflectances of each step. This leads to the quintic equation, due to him, that defines Value. He points out that the Munsell system is equally spaced along each of the three coordinates

Hue, Value, and Chroma but that it does not meet the requirement that equal distances in any direction be equally different, even when the three are properly evaluated with respect to each other by the Nickerson "fading index." This "fading index" (Nickerson 1936) takes into account the increase in circumference of the Chroma circles with increasing Chroma and also adjusts the three scales for Hue, Value, and Chroma so that equal distances on each represent equal perceptual distances.

He points out that the failure of the Nickerson index is due to two facts that it does not take into account. The first of these is the fact that as Chroma increases Hue increases in importance at a different rate; that is, that Hue spacing is not independent of Chroma even after correction for the coordinates and that, starting at any stimulus, the rate of increase of perceived difference diminishes with distance; it is not linear with the distance.

Because of these effects, he considers the still further possibility, which he attributes to a suggestion made to him by Carl Foss, of having the background against which differences are judged always intermediate between the two samples, not only in reflectance, but also in chromaticity. In this way it is hoped that both effects will be present to the same extent in all equal distances. In effect, this introduction of an intermediate surround supplies the fourth dimension that our studies have indicated is required. It does this, interestingly enough, by reducing small color difference judgments to the case of stimuli *seen as nearly as possible at* G_0. We saw from our study of chromatic surrounds that departures from the chromaticity of the surround showed a maximum rate of change in the neighborhood of the surround chromaticity and luminance. The suggestion, therefore, *maximizes* all small color differences and at the same time uses the Munsell system to produce the gamut of the G_0 colors. In effect it makes the Munsell stimuli act like a constant-brilliance set but includes departures from this set in the color difference, thus including all four variables in that difference. Judd concludes that Ideal Color Space so defined is probably the best that can be done with present knowledge and states "to my knowledge, the facts of vision so far established experimentally do not contradict this idea. It is possible that Munsell color space is a good approximation to ideal color space redefined in this way." The difficulty with the suggestion from the standpoint of the perceptual variables is that while it supplies a possible mechanism for the quantification of small differences based on the Munsell scales, it does not clearly identify the variables themselves.

With these various considerations of the way the perceptual variables of color might be portrayed and scaled, I shall leave this phase of the subject. However, the effect, just mentioned, of the increasing perceptual spacing of hues with increasing purity raises again the question of the relationship of hue to brilliance and, more broadly, the relationship of colorimetry to the perceptual variables. Although not enough is known for definite conclusions, it may be well to review the problem briefly in the next chapter before changing to the very different phases of color perception treated in the next section.

Ten

COLORIMETRY AND THE PERCEPTUAL VARIABLES

We have, perhaps, given sufficient consideration to the way in which luminance relates to brightness and lightness, but the manner in which the basic color-mixture data—chromaticity diagram and the derived functions of colorimetric purity, dominant wavelength, and the like—relate to hue, brilliance, and saturation has been found to be, at best, obscure. Some of this obscurity can be decreased by examining further the way in which some of the data have been obtained. It seems not possible at present to establish by such an analysis a complete correlation between the Standard Observer and these three variables. We can show why this is difficult, if not impossible, to do from tristimulus colorimetry, and also suggest what further data are needed, however they may be obtained. To do this I shall make full use of the CIE Chromaticity Diagram as a descriptive technique to *illustrate the operations* I shall be discussing, thus taking advantage of the fact that all possible physical stimuli are represented on or within the spectrum locus of this diagram. The major properties of this diagram that I shall be using are the facts that additive mixtures of any two stimuli are represented by points on the straight line connecting the two and the corollary, that stimuli along any line through a given "white point" are complementary if on opposite sides of it. With these facts in mind we can use the diagram to illustrate the operation by which the basic "color-mixture curves" are obtained and analyze their meaning in somewhat different terms than is customary.

TRICHROMATIC COLOR-MIXTURE DATA

The CIE color-mixture data are derived from a combination of data obtained by Wright and by Guild. For present purposes I shall discuss the subject in terms of hypothetical experiments that are equivalent to those actually performed but based on the combined results. The basic experiment is in two distinct parts: the matching of an arbitrarily chosen achromatic stimulus by a mixture of three selected monochromatic stimuli and the production of metameric pairs of stimuli by mixtures of these with monochromatic wavelengths throughout the spectrum. I shall consider each of these experiments separately.

The three monochromatic primaries and the chosen achromatic point are shown on the diagram of Figure 10-1. The primaries have wavelengths of 700 nm (R), 546.1 nm (G) and 435.8 nm (B), and the achromatic point is that for an equal-energy source (W). When these primaries are mixed to produce a metamer for the white, the luminances of R, G, and B stand in the relation 1:4.591:0.060 respectively. By considering the diagram we can state somewhat more accurately just what this means.

In such a match the final result is independent of the actual procedure used to reach it, so we are free to consider any systematic technique by which it might have been accomplished. In effect the ratio of the luminances of any pair of the primaries, say B/G, is adjusted until the hue is complementary to that of the third; the third primary is then added until the complementary mixture is just neutralized to produce the achromatic chromaticity. A luminance match is then produced by changing either the total of the primaries or the achromatic comparison stimulus, an operation which does not change the *ratios* of the primaries. On the diagram, the chromaticity R_c is produced by B and G and then R added until the chromaticity W is reached. The final result can also be considered to have been reached by way of G/R followed by B and by R/B followed by G.

It is apparent, from this approach, that the sum of the luminances of the primaries equals that of the white. The individual luminances, however, are the amounts of each individual primary necessary to neutralize the complementary mixture of the other two. Their widely different values indicate clearly two quite separate phenomena, the additivity of luminance and the cancellation of hue. Furthermore, in this latter effect the three primaries differ in what we can again call, for want of a better term, their "strengths," but which we cannot at this point identify with our "chromatic strengths," even though they are obviously similar in nature.

At this point, these values are taken as defining the units for the three primaries, both for convenience because of the great disparity

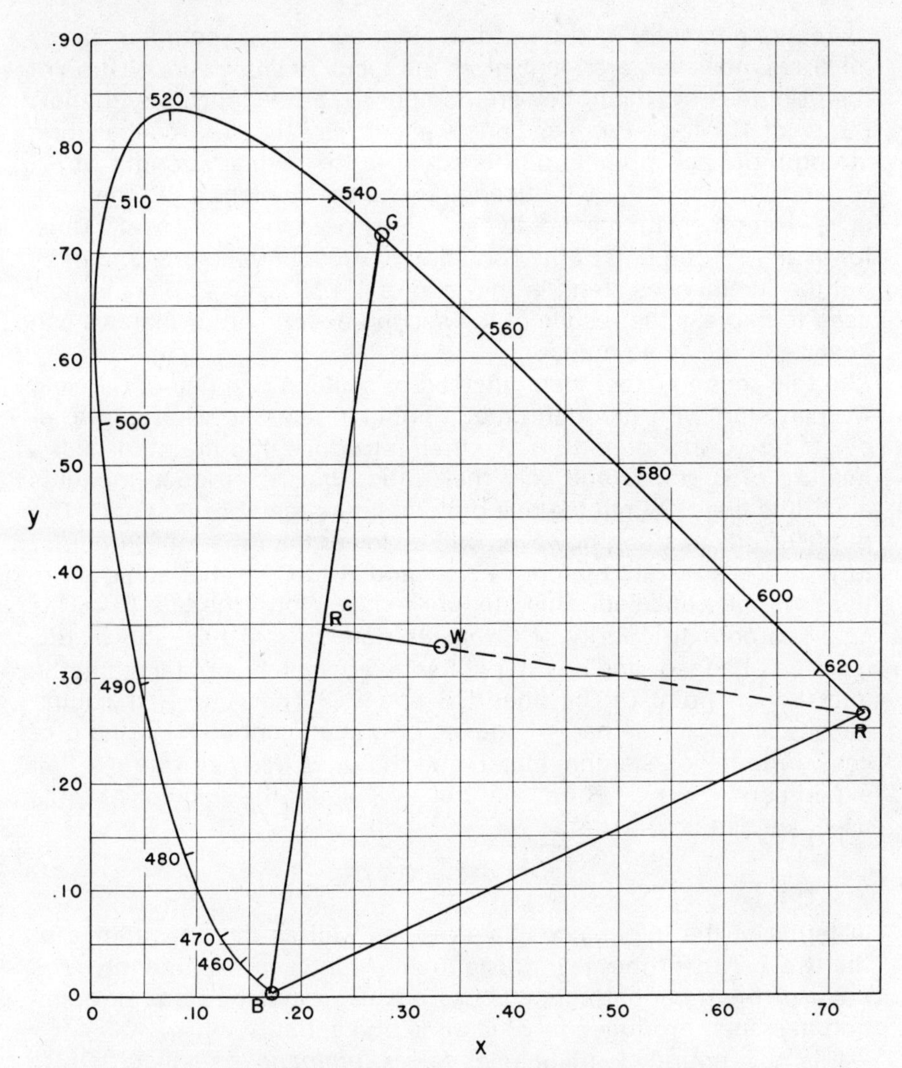

Fig. 10-1. Formation of white (W) by CIE basic primaries R, G, and B.

in luminances involved and because, with respect to white, these amounts of the primaries are in some sense equal. Thus R, G, and B are said to be equal to each other when their luminances stand in these relations to each other. It is apparent that each unit can thus be thought of as consisting of a luminance factor multiplied by a strength factor. Each product (unit) thus balances, for total strength, the complementary mixture of the other two. These complementary mixtures, however, are themselves the result of some cancellation of the primaries of which they are composed, as evidenced by the low purity of R_c, for example. Thus we can say that the cancellation strength of each primary unit is equal to the similar strength of R_c, G_c, and B_c, respectively, but can draw no conclusions as to their relative strengths with respect to each other or to the white itself. Thus, for example, one unit of the R primary corresponds to one unit of G, but the luminances stand in the ratio of 1:4.59. These units are now used to express the results of the second experiment, which we can analyze in the same manner.

In this second experiment, instead of white in one half of the field we can start with monochromatic light of, say, the wavelength indicated by C in Figure 10-2. We then introduce into the other half a mixture of B and G and vary their ratio; that is, produce mixtures along the line BG until the hue of the mixture matches that of C. The purity of this mixture, however, will be lower than that of C. In order to produce an exact match we now add R to C so that some of its hues will be canceled. This mixture moves along the line RC. This changes *both* the purity of C *and* its hue, so that the ratio of B/G must be changed, and so on, until an exact match is obtained at the intersection point of the lines BG and RC. The sum of the luminances on each side has, of course, also been kept equal in the process. In terms of the final match and the units derived from the first experiment, if we let R be one unit of R and *r* the number of units used, we can now write:

$$bB+gG=cC+rR$$

In terms of the luminances involved, as well as the appearance of the two halves of the field, this equation represents an identity. The color of the monochromatic C has not been matched but an equation has been produced in which it is one variable.

It is now entirely legitimate, *as far as luminance is concerned,* to transpose the *r*R term to the other side of the equation and write:

$$bB+gG-rR=cC$$

If this procedure is carried out for all wavelengths, using the appropriate primaries, we obtain an equation for each wavelength in terms of the three primaries. These equations then become the basis, again by use of the fact of additivity of luminances, for the

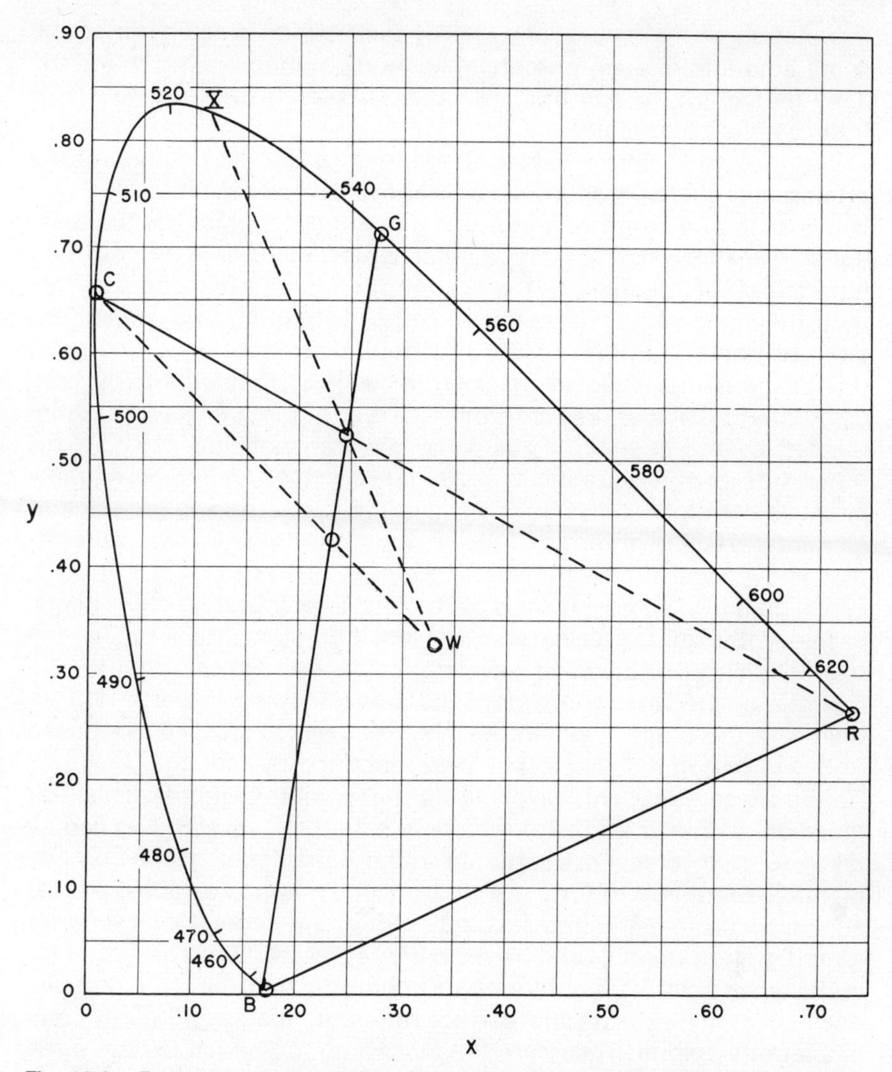

Fig. 10-2. Formation of metameric colors by mixtures of Band G and Band C for color mixture curve data. Dotted lines show shift of dominant wavelength from C to X.

calculation of metamers, any energy distribution calculating to the same amounts of each primary in such an equation being found exactly metameric to any other. This is, in sketch form, the basis of trichromatic colorimetry.

We need to inquire what has happened to the "strengths" of the primaries in this operation. We saw above that each unit R, G, and B is made up of a strength factor and a multiplier that makes the units equal to each other for a particular white mixture. We know very little about the meaning of these strength factors *except* for the fact that they are *vector* quantities. Thus in forming the white, the strength factor of one primary has neutralized that of the complementary mixture while in the mixture with C, R has *partially* canceled the hue and shifted it to the different dominant wavelength indicated by the line WX. Thus while the transposition of *r*R to the other side of the equation is legitimate for the luminance equality because it simply subtracts the same luminance from both sides, it is uninterpretable in terms of the vector strength factors without further knowledge of how these interact.

We do have some knowledge of this interaction in the special cases of the two wavelengths that are complementary to the primaries R and B. For these wavelengths we can assume that the directions of the vectors are opposite along the line through W. Furthermore the units in which the primaries are expressed are based on these directions, since they were obtained by reducing mixtures having these dominant wavelengths to the white stimulus. Thus, for these pairs, the strengths of the complementary wavelengths can be taken to be inversely proportional to the luminances of the complementaries required to form the white. Since it is well established that such a trichromatic analysis would hold for any three such primaries and that all of them would be properly represented by the CIE chromaticity diagram, it is thus not surprising to find that this diagram does correctly indicate the correct ratios for the strengths of complementary stimuli. These are the MacAdam moments. On the other hand it is equally *not* surprising that these strength ratios fail to properly indicate the relative strengths of wavelengths that are *not* complementary, since each would relate to a different set of primaries.

We have also seen that for complementary wavelengths our chromatic strengths, when measured against the same white, indicate these same ratios. Our strengths, however, are all individually measured against white and so *do* properly indicate their individual strengths on a common basis. It seems apparent that to obtain this result based on the CIE data we would need some law of combination of the vector strengths, or at least some way of relating them to each other. It seems unlikely that these can be deduced from the CIE data alone, or even from the original data before the many transformations. Note that when two noncomplementary wavelengths are mixed there is not only a partial cancellation of the hues with production of an equivalent amount of white but also a shift of the mixture to a new dominant wavelength, that is, a distinct shift in hue.

OPPONENT HUES

The most basic study we have of the hue shifts involved in mixtures of monochromatic stimuli is that of Hurvich and Jameson outlined in Chapter 7. In this work each of them determined, for his own eye, the "absolute" white (c. 5500 K for each), the luminous efficiency function, and the four spectral stimuli corresponding to the unique hues. These four stimuli (for J) are shown in Figure 10-3 at R, Y, G, and B and their white at W. Note that these five stimuli were chosen by inspection based solely on hue (or its absence for W); Y is neither reddish nor greenish and so on. It is also important that these stimuli (except for W) were chosen in the presence of a surround of their W, with both stimuli at approximately 10 mL.

Hering had pointed out that these hues cancel each other in the pairs B-Y and G-R. Thus if we have a greenish-yellow stimulus and gradually add to it unitary red, the green component of the color decreases until it disappears, leaving unitary yellow, and then becomes reddish yellow. No trace of redness is seen until after the greenness has disappeared. Similarly, if unitary blue is added to the greenish-yellow stimulus, the yellowness first disappears, leaving unitary green, and then blueness appears. Any stimulus, monochromatic or otherwise can thus be characterized by the amounts (luminances) of the two (or one) opposite unitary hues that have to be added to produce these cancellations. The end point, in each case, is the visually recognized disappearance of the opposite hue component; there is no matching operation involved.

While B and Y are roughly complementary to their W, as can be seen from the diagram, R and G are not. It follows from this that while mixtures of B and Y by themselves do not produce appreciable amounts of R-G response, this is not true of mixtures of R and G

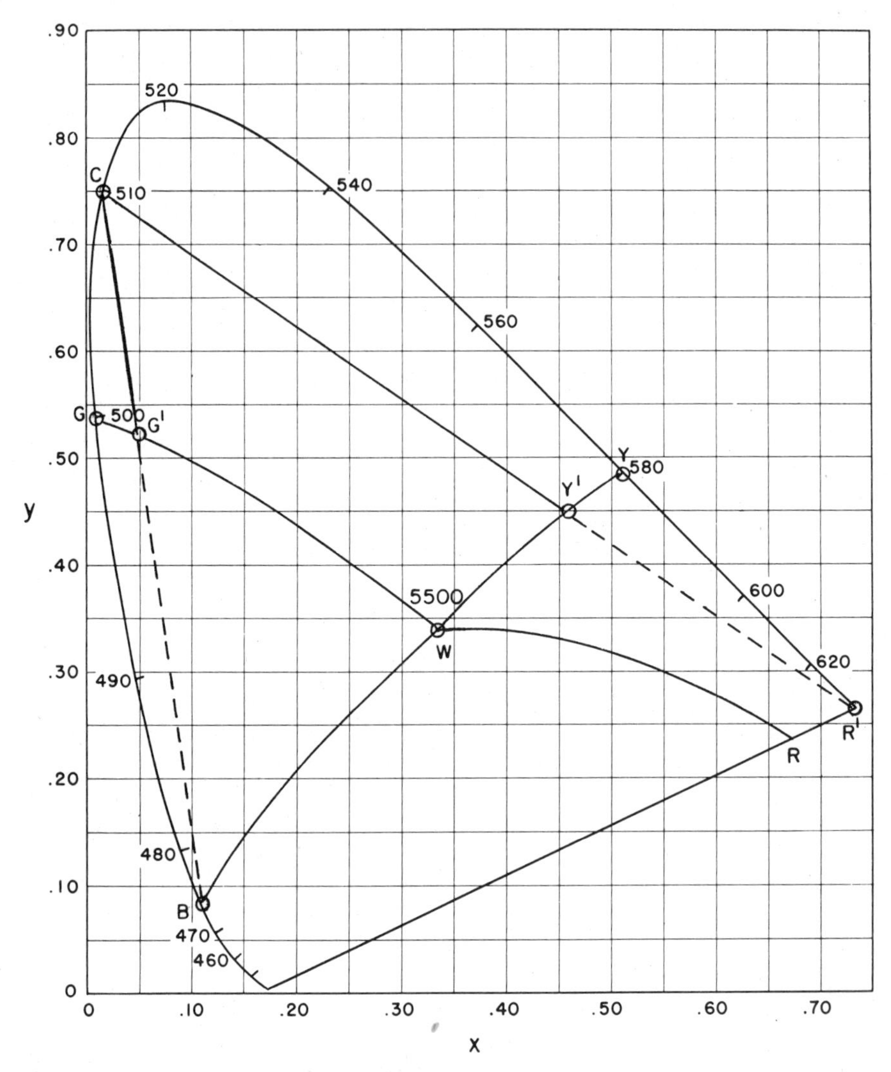

Fig. 10-3. R, Y, G, and B are the unique hue primaries (obs, D.J.); W, the white point; W-R, W-G, and so on, schematic lines of constant hue; Y′ is the unique mixture of C with R′ (see text) and G′ that of C with B.

by themselves. These always have a yellow component, equal-part mixtures of these giving unique Y but with a strong admixture of white. It also follows that the canceling stimulus used does not have to be exactly a unitary hue since, for example, a slightly yellowish red would still, through its red component alone, cancel green, but the yellow component would add slightly to the residual yellow. They took advantage of this fact to simplify their experiments by using a monochromatic red (R′) at 700 nm rather than the red and blue mixture necessary to get the unique red (R).

This cancellation operation was carried out by each observer for monochromatic stimuli throughout the spectrum. Each wavelength could thus be characterized by the luminances of the opposite primaries found necessary to produce cancellation. These operations were carried out separately; the mixture was not carried through to white. In order to illustrate these operations on the CIE diagram it is necessary to know the chromaticity loci of the stimuli that give constant unitary hues with varying purity. These were not determined, or at least not published. From the Munsell system, however, we know that they are not straight lines, except perhaps for WY, and I have indicated them on the diagram of Figure 10-3 by arbitrary curves simply for descriptive purposes and to call attention to the fact that they are not straight; they have no other validity.

Descriptively, then, for a stimulus of wavelength C (a yellowish-green), B is added, keeping the combined luminance constant, until the mixture falls on line WG (at G′), and the result is recorded as the amount of B. Starting over, R is added until the mixture falls on line WY (at Y′) and the result recorded as amount of R. These operations are repeated throughout the spectrum, using the appropriate primaries. In this approach no metamers are produced and the relative amounts of the four primaries required to produce a white are not determined. The relative amount of each primary necessary to completely cancel its opponent hue at this intensity was, however, determined. Using these latter ratios, they then converted the recorded amounts of the canceling primaries (B and R in our example above) to the equivalent amounts of the hue canceled (Y and G).

We can perhaps gain a better insight into the factors involved in this gauging of the spectrum if we reconsider this whole operation in the same terms we used for trichromatic colorimetry earlier. Each of the four unique primaries may be thought of as having a different strength per unit luminance and, as before, these are to be considered as vector quantities. This time, however, we can say that these vectors have known directions, at least for the conditions of the measurements. The R and G vectors are opposite in direction to each other long the line R-G, and their strengths are inversely proportional (since the units are considered equal) to their relative luminances for complete cancellation. The same statement holds also for the strengths of B & Y and the units in which they are expressed. These data, however, do not establish a relation between the units of the R-G pair and those of the B-Y pair.

It is a tenet of the Hering theories that R-G and B-Y represent separate receptor mechanisms and that the sensitivities of these receptors with respect to each other are not fixed but are determined by the situation. (This assumption is based on hue shifts with intensity, adaptation effects, etc.) To establish this relation for the conditions of their experiments, therefore, Hurvich and Jameson made use of a further, visually recognizable set of hues, the equal-part mixtures of the adjacent unitary hues. If the wavelength representing unitary green is changed continuously until yellow is reached, it passes through a wavelength at which the hue is seen to change from predominately green to predominately yellow; and similar wavelengths can be found for the other pairs. Having determined these wavelengths for their primaries, they then adjusted all their data by multiplying all the results for one receptor by a constant that equalized the numerical values for these wavelengths with those from the other receptor. (Actually they simply moved their curves, plotted in logarithmic coordinates, until they intersected at these wavelengths.) They thus adjusted the relative strengths of their vector units so that they were equal at these wavelengths although, of course, at "right angles" to each other. Note that this adjustment by a single multiplying factor is an assumption based on the assumed independence of the mechanisms, that is, it was not established experimentally that the "constant" was not in fact a wavelength-dependent variable.

It is a further tenet of the Hering theory that there is a third receptor system, wholly independent of these, that responds separately to the luminances (brightnesses) involved and that the R-G, and B-Y receptors give only hue responses. In a previous study, they had each determined their own relative luminous efficiency functions. Because they considered that this represented a separate mechanism they now assumed that the ratio of the *sum* of the two chromatic responses (in the adjusted units) to the relative luminance of that wavelength was a direct measure of the "saturation" of that wavelength. Thus, in our example, they took the sum of the adjusted values of B and R, divided by the relative luminance of the monochromatic stimulus C (510 nm), as a direct measure of the "saturation" of that wavelength, and so on for the other wavelengths. They then showed that the logarithm of the reciprocal of these values as a function of wavelength possessed all the characteristic features of the published functions for the colorimetric purity threshold; that is, that it plotted parallel to the published curves, well within the wide variation of these curves.

It is difficult to justify this operation except in terms of the Hering theories. Superficially the chromatic excitations, which are expressed in (adjusted) luminance units, are divided by the relative luminances of the wavelengths and the result considered as equivalent to colorimetric purity, which would be the first divided by the *sum* of the two. If, however, quite aside from the theories, these chromatic excitations (valences) are considered not to be luminances at all but as numbers linearly related to the "strengths" of the primaries involved and that all the luminance is represented by the luminous efficiency function, then this ratio becomes an expression of the strength per unit luminance as a function of wavelength.

Since we found that our chromatic strength (S_λ) was similarly related to these same purity threshold functions, it follows that their results are also related to ours by a roughly constant multiplying factor. We are perhaps thus justified in assuming that their "chromatic valences," when added for a given wavelength, would be identical to a constant multiple of our S_λ values, as measured against the same white surround. (Note that our surround was c. 7000 K and 100 mL whereas theirs was c. 5500 K and 10 mL.) If we assume that this is so, then their results can be thought of as *analyzing our* G_0 *function* into hue components; that is, our function represents a total strength relative to white, and their results analyze it into the two hue strengths of which it is composed. Since they had shown as well that the relative amounts of these components also describe the hues seen (necessarily for at least eight points in the spectrum), the two studies together demonstrate a direct relationship of the wavelength of monochromatic stimuli to both hue and brilliance relative to a given white. Furthermore, if the adjusted vectors can be considered to apply to all cases, a mechanism is thus suggested for the calculation of the hue of any chromaticity for this white. Given the S_λ for this dominant wavelength we have seen that we can also calculate the brilliance and the saturation, thus making all five perceptions calculable, but only for these conditions. There remains the problem of changing from one white to another, but this could be accomplished if a proven relationship could be found between these studies and the CIE Standard Observer functions.

Hurvich and Jameson did find in the literature a treatment of the CIE data that, they were able to show, leads directly to functions so similar to theirs that they assumed the identity and went on to use it in place of theirs in the derivation of explanations of other known visual effects. To consider this treatment in some detail seems worth the digression for the light that it may throw on any eventual solution to the problem. It is found, however, that this treatment involves an arbitrary multiplying factor, again considered constant, and based on another theory of color vision. In effect the theory is the same as Hering's without the awkwardness of his "black-white" receptor (both are "zone" theories) but identifies the R-G response with CIE X and the B-Y with Z. The constant, however, comes this time from Munsell system Chroma rather than from equal-part hues.

THE ADAMS EQUATIONS

S. M. Newhall (1940), in a report of a study of the spacing of the Munsell system by a subcommittee of the Optical Society of America, pointed out that if 3(X-Y) was plotted against (Z-Y) for each of the then standard Munsell colors for a single Value plane, they plotted as nearly concentric circles, spaced fairly uniformly for Chroma. This possibility had been suggested by a theory of color vision published by E. Q. Adams (1923). Adams in a later article (1943) suggested a modified factor of 2.5 rather than 3 for the (X-Y) variable. This factor was later incorporated into equations for the Hering, Adams, and Muller theories by Judd (1951). Hurvich and Jameson (1955) considered the similarity of their chromatic valence curves to those given by Judd's equation, assumed the identity, and adopted the equations

$$y - b = 0.4Y - 0.4Z$$
$$r - g = 1.0X - 1.0Y$$

where y, b, r, and g represent the four chromatic valences now derived from CIE. Using these equations they were able to derive a curve, now based on CIE data, that was similar to the purity threshold curve. It is apparent, then, that by using their concepts and an arbitrary constant based this time on the circularity of the Munsell chroma circles rather than the equal-part hues, they demonstrated a connection with the CIE system; hence also, *to the extent that they are the same,* with our G_0 function.

Adams pointed out the arbitrary nature of his constant, and Hurvich and Jameson consider it as necessarily variable to account for hue shifts with intensity. Furthermore, its constancy with wavelength is an assumption of both theories. This formulation, therefore, is hardly a basis on which to build a system for the prediction of the perceptual variables. However, since the unique-hue approach, at least when coupled with the opponent-hue concept, does lead both to hue-naming predictions and to a chromatic strength function, and we have shown that this in turn can predict brilliance and saturation, the whole approach is very suggestive.

Perhaps it is not extravagant to conclude that something like our brilliance function, broken down into functions that will predict hue, may eventually be developed into such a system. At any rate it seems clear that one or the other or both must be added as standardized data to the CIE system unless a way can be found to derive them from the original data of Wright and Guild.

If we can assume that such functions can either be derived or established, it may be appropriate to recount briefly the problems in developing a workable system of what might be called perceptual colorimetry to distinguish it from the present stimulus colorimetry. From the spectral distribution of the energy reaching the eye from a given stimulus it is now possible, given a stated white, to determine the chromaticity and luminance of that stimulus for the Standard Observer. From these data we can derive the dominant wavelength and colorimetric purity of its monochromatic-plus-white metamer. If we knew the specific chromatic strength (S_λ) for this wavelength for that white, we could calculate the brilliance of the stimulus against that white and also its saturation. Given also the unitary hue vectors for this point we should also be able to calculate the direction of its hue vector, that is, its hue in terms of the ratio of the single components from each of the unique hue pairs. We could thus describe, quantitatively, all five color perceptions for the stimulus under these conditions. Such a goal is obviously attainable but of limited utility except perhaps with respect to color vision theory. In order for the calculation to be generally meaningful a number of other possibilities must be available. For example, we need to be able to predict the effect of changing to a different chromaticity for the reference white. We can now do this up through dominant wavelength and purity, which both change, but we know that S_λ for that wavelength has also changed as have, probably, the relative vector strengths and, perhaps, the directions for the unique hues. We thus need a basic principle for conversion from one reference white to another and this must include, also, shifts in the relative magnitudes of the hue vectors with relative luminance. It is conceivable, however, that such a principle could be generalized to include *any* reference chromaticity and thus include the case of a stimulus seen against any surround; in fact this has to be the ultimate goal for this phase of the subject. Even if this goal is reached, however, it would appear that the appearance of *both* the stimulus and its surround may require further consideration based on the adaptation of the observer if this is due to other circumstances.

This need not be considered a discouraging analysis; even the ability to calculate the five perceptions for a single condition would be a great advance over our present situation. It highlights, however, the apparent futility of any attempt to construct a chromaticity diagram, based on trichromatic stimulus colorimetry alone, that will be perceptually uniform in its spacing even for one condition in which more than one stimulus is present.

SECTION THREE

THE OBSERVER AND THE GENERAL STIMULUS

Eleven
FRAMES OF REFERENCE

In the first two sections we have considered stimuli only in terms of an observer with determinate physiological responses. We have done this in terms of stimuli of varying complexity, increasing from the single isolated stimulus up to the case of three closely related stimuli, and have at least implied that perhaps all possible complications can somehow be encompassed within this range. The observer has been considered throughout as a sort of necessary nuisance to which we have to turn for facts but is often found to be discouragingly variable. The observer, as a person, has hardly entered into the subject at all, except as regards the semantic problem of describing what he sees and the general difficulty of communication concerning subjective responses. At no point is the observer assumed to have any interests in the stimuli other than those involved in the subjective appraisal of the color perceptions that the stimuli produce. It is obvious that this is a rather special attitude, readily adopted by anyone, but it is not one likely to be adopted for any other purpose. Even if employed to accept or reject merchandise on the basis of color alone, an observer's attitude is quite different. It seems pointless, therefore, to extend the complication of the stimulus so that it stimulates the real world without first inquiring whether or not his attitude makes a difference in the colors that he perceives. Furthermore we have placed a limitation on the stimulus itself that is independent of the number and spatial distribution of the other stimuli present: we have specified that the area under consideration, *the* stimulus, be uniform over its entire area. This means that it has no gradients in its psychophysical variables over the area and that it terminates abruptly at its boundaries. We need, thus, question the effects of violating this and other assumptions. Perhaps most important, if we are to consider color perception at all broadly, we need some insight into the relative importance of color to an observer, not so much in respect to the world in general, but in comparison with the other aspects of appearance that we so rigidly excluded at the start, such as glossiness, transparency, and roughness. Only after considering such matters will it be possible to decide on the extent to which we would be justified in the pursuit of more complex stimuli. I might point out, however, that all this is true only because our subject is perception and not the theory of color vision. For that subject much of this section, at least in its physiological aspects, will be found irrelevant.

One consequence of the stimulus-response approach to perception is the belief that the response to a given configuration is unique. When a large number of observers report sufficiently similar perceptions to justify averaging the results, we tend to believe that a fixed relation of observers to this stimulus exists and tend to forget that we have also prescribed the context in which the stimuli have been observed. Specifically we unconsciously adopt from physics and mechanics the *unitary* nature of the outcome of a known system of causes. In doing this we err, not in the belief in cause and effect, which there is no need to question, nor in the belief that there can only be a single outcome, given sufficient knowledge. The error comes in the belief that what is seen must be a single entity. This is demonstrably false in perception and not a necessary concept even in mechanics. Suppose, for example, that we see a newspaper delivery boy ride down the sidewalk and toss a newspaper onto a customer's porch. The path followed by the newspaper is a single unitary curve that can be seen by an observer watching the operation. What the observer is more likely to see, however, is the boy in motion along the sidewalk and the paper moving from the boy to the porch; in other words, what is, from the standpoint of mechanics, a single motion due to two forces is *seen* as two motions because these are the more obvious facts. Note that the observer, if he *looks* for it, can see the unitary motion of the paper but that this requires that he have a different attitude, one which is, in fact, quite analogous to what we require in our studies of visual response.

When an observer sees a color stimulus that is obviously produced physically by two separate causes acting on the same spot, he is more likely to see two different colors, one due to each cause, than he is to see the combined stimulus resulting from the two, *unless* the combined stimulus is so startlingly different from either that it attracts his attention. At the risk of carrying the analogy too far, suppose, in the case of the newsboy, that there is a flagpole next to the walk leading to the porch and that the curve taken by the paper as thrown makes it collide with this pole. The observer is then forced to see this unitary curved path. Specifically to our point, if an observer sees blue light falling on part of a green object and the result at that point is red, he will see the unitary red stimulus. If the result is a more normal bluish-green he will see a green object in blue light. There is nothing esoteric about it; this is *normal* perception. Again, by a deliberate change in attitude he can see in this latter case the blue-green of the combined stimulus without difficulty.

FRAMES OF REFERENCE

There is, therefore, nothing metaphysical about the statement that when an observer looks at a natural scene he sees the illumination as separate from the objects being illuminated. Under normal conditions an observer looks at situations, not at stimuli as such. His perceptions will, however, be strongly affected by the extent to which the true situation, from the physical standpoint, *is* apparent to him *and* by the attitude that he takes toward them. I want to introduce here a rather new concept which I shall call the "frame of reference" in which color stimuli are seen. The purpose of the rather lengthy prologue has been to point out that by this term I do not imply a change in the origin of coordinates of the system in which the color is seen but a change in the phase of the perceived color which is seen as applicable to the frame of reference. The distinction will become apparent as we proceed but an immediate example is the perception of a shadow as pertaining to the illumination frame of reference rather than to objects and the combination color as pertaining to yet another frame of reference that is stimuli as such. It is simply a more sophisticated phrasing of "what the observer is looking *for* and why."

On the other hand, the division of perception into frames of reference, while helpful, carries with it all the dangers of categorization and the attendant "either-or" fallacy. What is seen is not "either" illumination "or" objects but a sort of felt combination of both with the emphasis very much dependent on the circumstances. It is this fallacy that has so confused the literature on the so-called "brightness constancy" phenomenon (which we shall discuss presently) and that has made a perfectly straightforward subject seem so remote and mysterious.

With this warning in mind, I shall proceed to discuss the three major frames of reference pertinent to what an observer sees when he looks at a naturally occurring collection of stimuli, whether outdoors or inside. I shall call them "illumination," "object," and "stimuli," adding "as such" or *"per se"* to the latter as seems necessary. I shall classify them in terms of the physical or psychophysical stimuli involved and in terms of the five perceptual variables already deduced. The task is already simplified to the point where I can be quite brief.

ILLUMINATION PERCEPTION

Consider first the illumination frame of reference or, for short, illumination perception. Unless we see the actual light source itself, which would be seen as an isolated stimulus, the only evidence of the nature of the illumination in a scene comes to us from its effects on the objects on which it falls. Yet these effects are usually sufficiently numerous and different from what we have learned to expect from objects that we at once see the illumination directly. Perhaps the two major effects are the gradient in intensity (or lack of it) related to distance from the source and shape of objects and the nature and intensities of the shadows thrown by objects. Such effects are often called "clues" to the illumination. While this is convenient terminology in explanations of how the perception can occur, it implies a detective-like attitude on the part of the observer that doesn't accord with the fact that he simply *sees* the illumination. The amount of information that is conveyed to the observer by these two effects alone is often quite startling to the uninitiated. From them an observer is often able to state the size, shape, distance, direction, and number of the light sources involved, even though he can seen none of them directly. In fact the perceptions in this frame of reference are so necessary to an observer in his interpretation of a scene that if the effects are ambiguous or missing, as they often are in a poor photograph, he may see the scene "incorrectly."

Light seen as illumination displays all five color perceptions. In the usual situation it has little color and the predominant perceptions are brightness and direction. It is often stated in the literature that an observer is unable to judge the luminance level (strictly, "illuminance" level) of a scene because his eye sensitivity level adjusts to each level by adaptation. This is a rather misleading oversimplification of the facts due, in part at least, to confusion of the variables brightness and brilliance. Thus the argument usually runs that some scenes that on actual measurement are found to have low illuminance seem "as bright as daylight." Often the mere substitution of the word "brilliant" for "bright" will demonstrate the confusion.

The true situation is found to be quite complex, as is usual in perception. A distinction has first to be made between awareness and judgment. Thus the statement that on entering a room it seems quite dark (or bright) but that after a while it seems like any other illumination level must be interpreted in terms of both physiological and mental adaptation. A person not interested in the subject or not introspectively inclined, or neither, will readily accept the statement that after adaptation all levels look alike. As a matter of fact they do not, but even for a trained observer the precision of judgment is low. The range of adjustment of eye sensitivity to luminance within the photopic range is at least several thousand to one, while the range of brightnesses within any one level is at most 200 to 1 and more normally 40 to 1. The whole nature of the adaptation process seems to be an adjustment which maximizes this latter range for the subject being viewed, whether this involves the scene as a whole or some *part* of the scene which is of immediate interest. Adaptation must not be thought of as a static phenomenon; it is constantly and valuably dynamic. Looking at a bright part of a scene will darken it and looking at a dark part brighten it but only to the extent that doing so aids perception of detail and only within the physiologically limited range. Hence we see people shading the eyes, wearing dark glasses, and so on. The fact remains, however, that a person can judge the absolute illumination level, at least for order of magnitude, for all levels. What he cannot do is distinguish between the effects of illumination gradients and those of absolute level. Both a day with uniformly overcast sky and a room with completely uniform ceiling lighting and light walls look much darker than the same absolute levels when there is directional illumination present. This, however, is more appropriately discussed later under the subject of contrast.

The point here is that brightness is the major perceived variable of light seen as illumination. Except for directly visible light sources (or their reflections in shiny objects), this is the only aspect of any scene to which the term brightness is properly applied. Strictly speaking, there are no bright *objects* in a scene, only brightly lighted ones.

Perhaps the second most important perceived aspect of illumination is brilliance. We have seen that the brilliance of a stimulus depends on its luminance relative to the existing adaptation level and on the chromatic difference between the two. Levels much lower than the adaptation level produce the perception of blackness or grayness; those higher produce fluorence. These aspects with respect to illumination are seen in the blackness or grayness of shadows and the increased brilliance (often actually fluorence) of areas more brightly lighted than the rest. One reason that we do not see these areas as fluorent is that we think of fluorence as the perception of fluorescence and hence as an *object property.* An interesting illustration of the dependence on adaptation is the *increased* brilliance of fluorescent materials on "dark" days as compared to their appearance on sunny days. The explanation appears to be that dark days are usually overcast and hence have uniform illumination, whereas sunny days have a wide range of illumination intensities. Since adaptation is controlled more by the brighter parts of the scene than the darker, the fluorescent object has a higher luminance relative to adaptation on a dark day than on a bright one. The effect is often quite striking. On the other hand, once it is realized that fluorence is not the perception of an object property, it can be seen to be the correct description of many stimuli ordinarily simply called bright.

It is not immediately apparent what role, if any, the perception of lightness plays in illumination perception. As the perception of relative luminance, it is certainly present, yet there would appear to be little point in calling it a separate phenomenon. More accurately, there appears to be little *need* for a distinction between lightness and brightness with respect to illumination. As a matter of fact both words are somewhat ambiguous when used for relative light intensity. Thus it makes little difference whether we say one part of a room is lighter or say it is brighter than another, but to be clear in our meaning we may have to say it is more brightly lighted. It is probably better to think of lightness as playing a minor role in illumination perception and reserve the use of the word, as is usually done, for perception in the object frame of reference where it plays a major role; it is a mistake, however, to say that it is absent.

The perceptions of hue and saturation, representing as they do the qualitative and quantitative aspects of any chromatic perception that may be present, play pretty much the same role in all frames of reference. Hue is present or absent in the degree indicated by saturation and, from one point of view, is simply a matter of perceived cause. From an analytical point of view, it is just this fact that causes the apparent anomaly of two colors being perceived separately from a single stimulus. That the anomaly is only apparent and is *due to* the analytical approach is at once evident when it is realized that we *always* see at least two colors from any stimulus except in the very simplest and usually artificial situations. In the most ordinary of situations we say that the light illuminating the objects is colorless; we *see* that it is colorless in spite of the fact that every object in sight may appear colored. It is not anomalous to see the light as separate when it is yellow. It is only the analytical approach that says the light from a particular object has been modified and has become a new single stimulus; and it is, in the ordinary case, only the analytical expert who can see it that way. We shall return to this when we consider object color perception below, but point out now that one of the major errors in the application of colorimetric thinking to perception is the assumption (usually unconscious) that what is seen must be explicable by a simple combination of a single stimulus and an eye sensitivity modified by color adaptation. This combination *can* be seen but usually requires artificial aids.

From the standpoint of adaptation, on the other hand, it is usually the illumination that produces the major effect. To the extent that the modification of the light by the objects adds up to a neutral we can say that illumination is the only cause. Here again, however, we must think of adaptation as a dynamic process; and in a region of a scene where there is a predominance of one hue, there will be an adaptation shift tending to decrease this predominance and so amplify the local differences. Adaptation to both the luminance and the chromaticity characteristics of a scene must therefore be thought of as variable and tending to maximize the ease of perception of both brightness and chromaticness differences in the immediate areas of concern to the observer.

OBJECT COLOR PERCEPTION

We turn now to the perception of colors in the complex, object frame of reference. I have started with illumination perception, not because it is the most important, but because from an analytical point of view object color perception makes no sense unless illumination perception is understood. To the naive observer the object frame of reference may be the only one of which he is ever aware and then, not as a separate phenomenon, but as the whole of his visual experience.

We do not need to enlarge on this frame of reference because we have already done so in an earlier chapter without calling it that. In object color perception, brightness is not seen as due to the objects; and the number of variables, as we have seen, reduces to four. The major difficulty in the discussion of object color lies in the tendency to say that, at a given moment, an observer is perceiving in one way *or* the other; he seldom does this. An observer is always aware of the illumination although seldom consciously. He has simply trained himself to take such action, automatically, as will maximize his perception of the object properties. We see this in everyday commonplace actions. If a glossy surface reflects light into his eyes, he will move his head, change his position, or, if possible, pick up the object and hold it so the reflection disappears. If the light is too dim he will take an object to the window or to a light. If the illumination is too bright he may shade it with his hand. In other words, he deliberately manipulates the illumination to see the object color; they are two separate things to him but he is usually interested only in the object.

It is worthwhile to digress for a moment to consider this fact from two quite different points of view. Take first the matter of the glossy surface. It is one of the problems of spectrophotometry that a decision must be made on inclusion or exclusion from the measurements of the nonselective reflected light from the outer surface of materials. From the standpoint of perception (what the object will look like in the hand) there are clear-cut answers only for the cases of very high gloss and complete matte. High gloss is perceived in the illumination frame of reference and so should be completely excluded, *if possible,* although the extent to which it is difficult will also bother the observer. In the case of a completely matte surface nothing the observer can do will change it, and so it is not seen as separate; he sees only the combination color. There is also nothing the spectrophotometrist can do about it unless his sole interest lies in the selectivity below the surface, in which case he can eliminate the surface reflection (actually make it glossy) by wetting it with a medium of appropriate index of refraction. A layman often amuses himself at the shore by wetting stones for the same reason. For spectrophotometry the answer can only be that a measurement on a semi-matte surface must be made either for the intended viewing conditions *or* that the two measurements are made showing the *limits* of the variation.

The second point of view concerns the rather absurd discussion that crops up every now and then as to whether it is proper to say that an object "has" a color. What the object has is selective action on the light, but what the light has is selective energy distribution, both before and after reflection or transmission. Arguing about it is about as sensible as questions whether a horse is moving forward or backward or standing still when you pass him in an automobile. I bring up the subject because it leads to a very interesting and, I think, very important point in regard to color perception in general. There is an amazing discrepancy in the ability of any person to discriminate between two juxtaposed colors and his ability to pick a color from an array that will match one he has just seen separately. Even after fairly short delays his ability is quite poor and after considerable delay it can only be described as bad unless he is trained. The ratio of the number of colors he can discriminate easily to those he can positively identify *by themselves* is, at the very least, 1000 to 1 and for the quite naive it must approach 10,000 to 1. Even for fairly good observers, positive identification of a color without a comparison is of the order of 30 colors, and for the naive it may be more nearly 8 to 10. It seems to me that the reason for this is the same as the basis for the above argument. In everyday life the colors of objects are not stable and there is no point in trying to assign an exact color to an object; accordingly people do not *attempt* to train themselves in this respect. Saying that an object is white or green or brown is about as far as they can safely go in describing it. In artificial situations where colors are stable observers *can,* without too much difficulty, train themselves to recognize and name, according to some system, upwards of thousands of colors with quite good precision. It is not so much the lack of ability that makes the average person such a poor "judge" of color as the fact that he has found it unrewarding to try because of the frustration of the normal situation. This does not change either his belief, or in fact, his direct perception that the color he sees is a property of the object. He blames his failure, *quite correctly,* on the illumination, and at this point he is fully conscious of it as an entity quite separate from the object.

LIGHT SOURCES

In discussing the perception of illumination and of objects we have rather sidestepped the direct perception of light sources as such. Perceptually they form a special class of objects and normally can be characterized by the three variables of an isolated stimulus, hue, saturation, and brightness. I hesitate to speak of them as being seen in a separate frame of reference, however, because the distinction is often based on knowledge rather than perception, and this changes the applicable variables.

In the first place, it is necessary to make a distinction between the light source and the illumination. As used in physics the term "light source" usually refers to a self-luminous body radiating energy in the visible range and is usually thought of as the only, or at least the principal, cause of the light present. Perceptually it is thought of as the cause of the illumination, a concept quite different since it is probably safe to say that in the average scene a far greater area is illuminated by light reflected from other objects than from the principal source itself. In the second place a distinction must be made between self-luminous objects that are perceived to be a source of illumination and those that are perceived *not* to be. Finally, of course, we have to separate out illumination that is seen to be due to light arriving by way of reflection or transmission.

An observer usually finds the principal source of light in a scene an annoyance, and direct viewing is usually avoided as far as possible. The reason is obvious: if he looks directly at it his eye sensitivity drops, often to the point where he sees little else, and he experiences the phenomena that we call "glare" and "afterimages." It is, therefore, somewhat academic to say that it has the variables of an isolated stimulus. Most light sources that are intentionally directly perceived lie in the same range of luminances as the light from objects. What is seen will depend first on the luminance relationship to the adaptation level and second on whether or not the observer realizes, that is, *sees* that a particulr object is, in fact, self-luminous. There is a bewildering array of possibilities but we need to consider only a few typical examples.

The adaptation level (sensitivity) of the eye is set by a running average of the luminances and chromaticities of the scene as the observer's gaze darts around it. The perceptions of the various stimuli involved will depend on their relations, as stimuli, to this state. In terms of brilliance, for example, the amount of grayness or fluorence seen will depend on this perhaps momentary luminance relation. This relation holds whether or not the stimulus is, in fact, self-luminous. On the other hand it matters very much what he regards as the cause of the stimulus. Let's take a few cases.

Suppose he sees a "shiny" curved object with a concentrated "highlight" reflecting the light source. He discounts this completely and sees only the glossy surface; the highlight is referred to the illumination. If it is small enough it has no affect on his adaptation. Similar remarks apply to all specular reflections (unless he is looking in a mirror); he normally looks "through" them at the object properties. Consider next the self-luminous "neon" discharge tubes so much used in window display signs and the like. These are, essentially, stimuli of constant luminance, independent of the illuminance of the environment, and usually do not radiate enough light to be sources of illumination. For these reasons, among others, they are perceived as self-luminous. Their appearance, however, depends very heavily on both the ambient illumination and the background against which they are seen, following essentially the course predictable from our simple stimulus and surround, but with the surround the variable. They thus can produce all five perceptual variables (they often look grayish in direct sunlight), but since it is known that they are self-luminous the higher relative levels are thought of as brightness rather than fluorence. A similar situation occurs frequently with indicator lights such as those often found in elevators to indicate the floor or the direction of motion of the car and, with increasing frequency, on the dashboards of automobiles. The luminance of these is often comparable to or lower than the illumination but they are known to be self-luminous. They thus, very often, produce the perception of brilliance, showing true fluorence or grayness, but are usually thought of in terms of bright or dim, because fluorence is thought of as the perception of a property of a fluorescent surface. Finally, an object which "catches the sunlight" in an otherwise shaded scene may well produce all the perceptions characteristic of a self-luminous object yet be assigned, in perception, wholly to the illumination. The category of light source frame of reference is thus not useful in thinking about the subject except, perhaps, where it is seen as the source of the illumination. Incidentally, these and like examples also serve to point up the fallacy of making a distinction between aperture and object colors as far as color perception is concerned; it is both artificial and unnecessary except with respect to the stimulus frame of reference, which we will discuss presently.

We note here once again the real distinction that must be made between a source of light and a source of illumination. In any situation in which there is only one true source of light (sunlight, for example), any light that illuminates a shadow must reach it by reflection, or scattering, or the like. Such light is usually sufficiently like that from the main source but may cause perceptual problems if the color is different and the source not obvious to the observer. There are two special cases which have been brought to my attention by color photography and that illustrate the point nicely; both are outdoor scenes in sunlight. If a person is photographed with the sun shining on one side of the face only and the shadow is illuminated by the sunlight reflected from a red brick wall, this side of the face will reproduce as an unnatural dark red. If the wall is not included in the picture this redness will be very noticeable and may be quite objectionable because it is seen as an object property. If the wall is included in the picture the redness will be perceived as illumination and may be completely discounted, that is, not seen at all! A better known case is that of blue shadows in pictures, particularly on snow. If the sky is clear blue and there are no nearby reflecting surfaces, the light in such a shadow comes from the sky and so, as a stimulus, is essentially the same as the sky. The fact is, however, that unless the sky is unusually blue the shadow as seen directly by an observer looks gray and not blue. This problem itself was once proposed to the Optical Society of America by I. G. Priest under the title "Blue Sky and White Snow," (unfortunately published only in abstract (Priest 1926)), and caused considerable discussion at the time. The answer is straightforward but is interesting because the situation puts an analytical observer in a perceptual dilemma. In experience, shadows are gray and normally so perceived; in experience, likewise, snow is white and is so seen. Therefore, since the blue sky is not usually thought of as a light source and certainly not as a reflecting object, there is nowhere to assign the blue stimulus. It is interesting to look at such a shadow analytically. If the conditions are right, this blueness can actually be seen apparently shimmering in the *space* occupied by the shadow with the shadow itself a clear gray. If the sky is too blue, of course, the shadows look blue whether they "should" or not, and if it is not blue enough no trace of the blue stimulus can be seen. In a photograph, however, the color reference is the white border; the print is a colored *object,* and the shadows look blue if they were at all blue as a stimulus. I should add, perhaps, that some processes also tend to exaggerate blue relative to the other colors but this is not usually the cause.

We have here two cases in which perception of the cause of an effect changes the perceptions related to a direct stimulus. They illustrate somewhat dramatically the fact that is true of all normal perception: that the illumination is seen separately from the object properties. The stimulus is normally divided in perception into parts appropriate to the frames of reference in which it is seen. This is essentially all there is to the phenomenon of the so-called "brightness constancy" or, as Judd more properly called it, the "approximate lightness constancy" of objects. To the extent that the observer can perceive the illumination as separate and so disregard it, he can see the true properties of the objects. To the extent that he cannot, he will see the illumination as part of the object properties.

Although many ramifications have been introduced by experimenters, the basic phenomenon involved in lightness constancy is the perception that a white or gray in shadow is still seen as white or gray in spite of the fact that, as a stimulus, it may be sending less light to the eye than a darker gray in full illumination. This is anomalous only to those who believe that the stimulus, *per se,* controls the perceptions. The confusing part of it lies in the fact that the extent to which the true object reflectances are seen depends almost wholly on the circumstances, that is, on the *extent* of illumination perception. An interesting demonstration of this comes again from photography. Any photographic print, or reproduction, consists of a collection of reflection stimuli on paper and, since they are usually viewed in uniform illumination, they tend to be seen as such. The picture, however, usually represents a scene with nonuniform illumination, and so there is a conflict between the reflectance in the print of an illuminated object and the reflectance of the object as it would have been seen with full illumination perception. The appearance of the object in the print, therefore, depends on the extent to which perception of the illumination is possible *from the print itself*. We cannot enlarge on the subject here; the interested reader is referred to my study of the subject with J. Klute (Evans and Klute 1944) and my book on perception in photography (Evans 1959). Briefly, all lighting gradients in a print are amplified, giving the appearance of much greater lighting contrast than was seen in the subject; hence the need for "flash fill light." Any area of the subject in which it is not *possible* to see the illumination, for instance an unrelated background, must be separately lighted if its true reflectance is to be made evident. Finally the nature of the photographic medium will affect the illumination perception and hence the lightness perceptions. A scene photographed in black and white will show the least "constancy." The identical scene in color will show very much more, a projected picture still more, and stereo color the most. Restriction to a portion of the scene, however, may still give serious "failures" in such areas as unrelated backgrounds.

The above discussion of the perception of real scenes has concerned only the reflectance aspect of perceived objects; similar remarks apply to the chromatic aspects as well. The perceived color of the illumination will tend to be subtracted from the stimulus color of the objects and their "true" colors seen. A particular case in point, at which surprise is often voiced, is the perception of a white in colored illumination. In the object frame of reference and as far as the observer is concerned, there is always a white in any real scene. If there is not one actually present he unconsciously supplies it from the appearance of the objects or, if the lightest neutral object would normally be seen as a gray, he may see this as his white. In this latter case, this amount of gray is subtracted from all colors he relates to it. The effect occurs very readily in projected pictures in a dark room and is one of the reasons for their beauty. In such a situation the lightest, most neutral area is always seen as white, and the brillances of all other colors in the picture adjust themselves accordingly. Advantage may be taken of this effect by limiting the lightest area in a scene to a medium gray; this was done in the early days of color motion pictures by the Technicolor Corporation, and the result was quite extraordinary in the increase in brilliance of the resulting colors. It can also be carried too far and some colors, those of lipstick for example, made fluorent. The perception of a true white object in a scene therefore tends to be independent of illumination color or intensity; it is seen as the anchor-point, so to speak, of the object frame of reference in its vicinity, quite independent of its psychophysical variables as a stimulus.

STIMULUS PERCEPTION

This brings us to the third frame of reference customary in viewing a scene, that of seeing, or trying to see, the stimuli *as such* without regard to the situation. This subject has a most interesting history, on which we can only touch, because it is a basic problem for an artist who is trying to reproduce the *appearance* of a scene. The problem for such an artist is easily stated but the solution is not at all obvious. He has at his disposal a range of pigments that will normally be somewhat greater than the range of colored surfaces he will encounter in the scene. His problem, however, is not to reproduce these surfaces but to reproduce their appearances, including the appearance of the illumination. Even on a day when the lighting is quite diffuse, the ratio of the highest illumination to that in the open shadows will probably exceed 4 to 1. This will already put the scene beyond the range of his paints because he has to reproduce both reflectance and illumination by the use of reflectance alone. On a clear day the lighting ratio may become as high as 40 to 1 and is hopelessly beyond any attempt to reproduce the actual stimuli as far as relative luminances are concerned. To a considerable extent he can handle this part of the problem by a distinctly nonlinear compression of the luminance scale, using, if necessary, different scales in different parts of the picture; and a large part of the remarkable skill shown by some painters lies in a judicious use of such compressed and distorted scales. In this the painter in oils has a distinct advantage in that his glossy medium will produce much lower minimum reflectances and so show a luminance range that can be higher than 100 to 1. At best, however, the exact reproduction of this aspect of the scene is severely limited.

When it comes to the chromatic aspects of the scene, however, he is normally not only not limited but usually has a range available that is considerably in excess of the actual object colors in the scene. He can, therefore, and usually must, use this chromatic excess to help him produce illumination perception in the observer. In this he is greatly aided by a perceptual phenomenon so strong that it is in many ways responsible for the success of color photography, color television, and color reproduction in general. I encountered this phenomenon many years ago and named it the "consistency principle." As long as all of the chromatic *relations* within a given scene are consistently like those of the original scene, the colors will actually seem to be the same in both, unless, of course, direct point-for-point comparisons are made. The actual discrepancies can be enormous, providing there is *no* inconsistency. The reason for the phenomenon is perhaps the great superiority of color discrimination over color memory, discussed earlier. In addition, both sunlight and incandescent tungsten light are usually perceived as yellow; and yellow, as we have seen, is the hue most like white. The skillful artist can use these facts by blending his colors with yellow, often using pure yellow for the most brightly lighted objects, less pure yellow for less brightly lighted objects, and so on, giving a very adequate effect of illumination.

Aside from the illumination and the reflectance range, however, he is still confronted by the problem of putting on his canvas stimuli that, in their surroundings, will produce similar perceptions to those of the original objects. This means that to a certain extent he needs to reproduce the stimulus, that is, the combination of object and illumination colors; and the local illumination, more often than not, comes by reflection from nearby objects. Many artists pride themselves on their ability to do this but describe what they do as something quite different.

Ruskin, in his wonderful book *Elements of Drawing* (see Ruskin 1971), first published in 1857 and still valuable for its insights, felt that the problem of the artist was to reproduce exactly the color that was there; and although he states that not many people can see that the grass "is really yellow" where the sunlight falls on it, he invented, or at least propounded, a method by which the artist could see the true color. This method was to cut a hole in a piece of the same kind of paper being used for the drawing and to hold up the paper so that the illuminating light fell on it; the object could be seen through the hole. He remarks that most people will be surprised by what they see. This is, of course, the "reduction screen" made famous by Katz many years later in a wholly different context and which, I am afraid, has become a bit of a red herring in the subject of color perception. The color seen under these circumstances is the actual color stimulus, related directly to the illumination. The "reduction screen" places all stimuli on the same basis and, more importantly, completely wipes out both the illumination and the object frames of reference. It permits the artist to at least see what the true light stimulus looks like relative to the white and gives him a place to start in his reproduction. That he neither can nor should copy it exactly is beside our present point.

This raises the important matter of the relationship of an observer to a reproduction as compared to a real scene. In a reproduction he is presented with a simulation of the *mixed* stimuli from the real scene, but these are now actually object stimuli. The reproduction is a new object with colored areas and an illumination of its own. We are, in effect, dealing with what amounts to a still further frame of reference. In this frame of reference the observer accepts the proposition that the picture is a reproduction and so consciously *tries* to interpret what he sees, as though it were a real scene.

Whereas in the natural scene all his experience has taught him to discount the illumination in order to see object colors, or object colors to see illumination, he is now given object colors that represent both and asked to separate them. Furthermore, he is asked to discount the illumination on the picture itself. The remarkable ability that he has of seeing object colors, which is largely unconscious, now works against him, making him tend to see the mixed stimuli as such, at least as represented. While it was difficult in the real scene to see the mixed stimuli, in the pictorial reproduction it is difficult not to. The major failure in interpretation of a pictorial representation thus tends to be the attributing of the mixed stimuli to the *objects in the scene,* that is, a partial to complete failure of the perception of *scene* illumination. It follows, then, that the art of *both* the photographer and the painter lies in making this aspect of the scene so obvious to the viewer that the perception does not fail; the painter by such methods as exaggeration and local distortion and the photographer by control of scene lighting, subject matter, angle of view, shadows, and so on. The tendency in photography, however, will be for shadows to look too dark and in paintings for them to look too light because neither artist tends to see them clearly in the original scene and both fail to take sufficient steps to offset or represent them, respectively.

An interesting question in the case of photography is that of a criterion of excellence to be used in assessing the quality of a process, the setting of an aim toward which processes might be moved to obtain an improvement. In color photography, whether it is to be used as the final product or as an intermediate step for the graphic arts, there always exists a definite relation between the light stimuli in the subject and the final reproduction, however complex the intermediate steps may be. Furthermore, we can assume for present purposes that these relations cannot be changed from point to point within one picture to suit the subject, even though there is liberty to change the *overall* effect, or even the relations of relatively large areas by such techniques as hand or mechanical "dodging." Only by hand "retouching" does the photographer approach the freedom of the painter. Thus in essentially all pictures of natural scenes there is partial or complete failure of illumination perception.

The only criterion known by which the excellence of a photographic reproduction may be judged objectively is the extent to which it does or does not produce point-for-point metamers for the subject stimuli. Since it is axiomatically impossible to reproduce most subjects by this criterion, the problem, from the colorimetric standpoint, thus becomes one of finding the most desirable distortions. In the past this has been accomplished largely by skilled people, making visual judgments on large numbers of pictures of different subjects and manipulating the myriad variables. The results are already amazingly good when the required conditions of exposure and handling are met, and failure is far more often due to errors in these than to any fundamental fault of the method.

Still, for special application purposes where some limitations may be possible on subject matter, it would be very desirable to have a more satisfactory reproduction criterion. It is quite apparent, I think, that the problem is very similar to that of setting color tolerances on manufactured goods of any kind and that this problem cannot be solved until we can evaluate the importance and magnitude of brilliance departures in relation to hue, saturation, and lightness.

The requirement, therefore, would appear to be that the reproduction be degraded systematically and uniformly in *all* the stimulus attributes, but uniformly with respect to the perceptual rather than the stimulus variables. It differs from the tolerance problem because it is the rate of change of the perceptual variables with respect to each other that is most important, rather than the static difference between two colors. There is also the difference that brightness as a variable is now wholly irrelevant as far as the initial scene is concerned and enters only to the extent that the illumination level on the picture changes the variables. The reproduction problem thus involves the same four dimensions but cannot be directly related to the color tolerance specification problem that is, by comparison, relatively simple.

We thus see that an observer looking at either a real scene or a reproduction does not, in general, try to see the actual light stimuli in front of him but looks at the stimuli from the standpoint of one or another frame of reference, almost consciously excluding perceptions extraneous to his immediate interests. Within each frame of reference his color perceptions can be arrayed systematically along coordinates representing the variables we have discussed, but the existence of more than one frame of reference denies the existence of a one-to-one relation between the stimulus and the perceptual variables. This is an extension rather than a denial of colorimetry but it introduces concepts that go far beyond mathematical relationships in the generalized stimulus. They introduce the observer as a person.

In closing this too brief chapter on the role of the observer in color perception, we might note the position of color perception itself in the larger subject of color appearance. Any color perception in real life is accompanied by a number of appearance characteristics that we ruled out rather rigidly as outside the subject of color. To the observer in any given situation, however, these other characteristics are often of greater importance than the color. This whole question of the relative importance of the various factors involved in perception needs evaluation, although it concerns, of course, applied color rather than colorimetry. Neither color discrimination nor color preference is a measure of importance, and in any real situation it seems likely that it is the relationship of the color perceptions to the other characteristics and to the nature of the object itself that will really matter. Perhaps we should simply note that in the larger subject of appearance, color is, itself, simply one frame of reference.

Twelve *THE GENERALIZED STIMULUS AND OBSERVER*

We have been concerned with rather specialized stimuli and, until the previous chapter, have not considered the observer as a person. Since our subject is perception and this is, in practice, a highly personalized phenomenon, we need now to look at the subject quite broadly, if only to see how our rather narrow and specialized approach fits into real life. We cannot give it the space it needs but perhaps what is essentially a series of notes about some phases of the subject may be helpful. Again we divide it into three parts: the stimulus and the restrictions we have placed on it, the observer and the assumptions we have made about him, and the broader aspects of the observer's responses.

ASSUMPTION OF UNIFORMITY

In our discussions we have placed a restriction on all stimuli that comes close to being contrary to the laws of nature. This is the requirement that the stimulus be completely uniform over the area considered. I am sure that anyone who has attempted to produce by hand a reasonably large area in which no variation can be detected realizes the artificiality of the assumption. It is made, primarily, to simplify the semantics and to give meaning (which is often not justified) to our measurements. Many artists go so far as to say that there is no such thing as a uniform color area in nature and that such a color in a picture is lifeless, dull, and unpleasant. We need to consider it here, however, only with respect to the relation between the actual stimulus and what is seen.

The uniform change in lightness from one end to the other of a fairly large area, say 1 by 2 ft, of a single color will not, in general, be visible to an observer, even though the total difference is quite large. Cutting apart this rectangle and juxtaposing the ends often causes surprise. The reason for this fact probably lies in our customary habits of vision. In a room with painted walls, any nonuniformity which grades continuously is seen as a change in the illumination, the wall itself as a uniformly colored surface. Accordingly any area uniformly graded, particularly for lightness, may tend to be seen in the same way. The fact that a similar gradation of hue does not show such an effect but tends to be seen as such lends support to this point of view. There is considerable evidence, however, that there may also be a purely physiological phenomenon involved.

If a moderately dark gray spot, say 1 in. in diameter, decreases in density outward from the center to 0 at its boundary and is compared with a uniform-density spot that is the same at the center but has sharp boundaries, the latter is seen as *very* much darker, almost unbelievably so. There is here no question of illumination gradient. While this effect is very probably physiological, it is still curiously involved in our perceptions of shadows. It was discovered by Mach (see Ratliff 1965) over 100 years ago that if such a gradient decreases at a uniform rate with distance (first derivative constant or changing at a uniform rate) the gradient only is seen. If however, the *rate of change* is not uniform, that is, decreases and then increases again (change in sign of the second derivative), even though the gradient everywhere decreases, there will actually appear to be discontinuities in the gradient. Specifically, if the rate of change with distance slows down and then starts up again a light and a dark band (known as Mach bands) will be seen where this occurs. Carried to the extreme of a sharp break at the edge of a uniform area, the dark band lies wholly inside the area and at least contributes to the darkness seen. In fact this phenomenon and the invisibility of the "blind spot" led Walls (1954) to his delightfully unprovable "filling in" hypothesis, mentioned in an earlier chapter, which states that the darkness seen is *wholly due* to the edge gradient. We cannot pursue the subject beyond noting a few instances in which this effect plays a considerable role in color perception, but it is a vitally important fact of vision, one that plays a large part in form and detail perception. It is apparently true, incidentally, that this phenomenon is restricted to the luminance aspects of the stimulus. I have been unable to produce Mach bands by changes in hue or saturation in the absence of luminance differences, although I have not had the opportunity to rule out the possibility that large brilliance differences at the same luminance might cause them.

We note here, then, only two effects that gradation in the stimulus may have on color perception. A gradation in the stimulus may be confused with illumination and hence produce ambiguity in the perception. Almost all shadows are produced by light sources of finite size and so have at their edges a gradual change in luminance. In general, then, all shadows tend to be seen as much lighter than they would appear with a sharp boundary. In fact, illumination of objects by a very small (point) source of light produces a very strange situation of great visual contrast, with the shadows tending to be seen as objects and the glossiness of shiny objects tremendously enhanced. It is for the same reason that in photometry a much higher precision is obtained by a sharp than by a diffuse separation line between two fields; and we can generalize to the effect that brilliance differences, at least those due to luminance, are heavily dependent on the nature of the boundary between two areas. In one experiment, however, we threw the dividing line of our instrument field out of focus and obtained the same G_0 values; thus it can be assumed that this effect simply increases the sensitivity of such settings, as might be expected.

CONTRAST

These phenomena lead us into one of the many phases of color perception that need reinvestigation in the light of the earlier chapters. The word "contrast," as it is ordinarily used, means a perceived difference between stimuli but is also used to describe the enhancement of this difference by adjacency of stimuli in space or time. It is used, furthermore, to describe not only the overall effect of a whole collection of stimuli but also the perceived rate of change of such differences with a variable, such as reflectance. There is thus some ambiguity in the use of the word itself although usually little doubt as to want is meant. The difficulty comes when an attempt is made to assign a cause to the perception, largely because it is assumed to be a simpler phenomenon than it really is. If we restrict ourselves to the magnitude of perceived differences there is no ambiguity, but we must not assume that there is a single assignable cause. *Any* perceived difference produces contrast but it is not necessarily cumulative from all the causes, and the different *kinds* of perceived difference are not necessarily comparable. In one sense, this is the underlying problem, in the difficulty of setting tolerance limits in color matching. It is not really possible to equate a contrast due, say, entirely to hue, with one due entirely to lightness because they are not only different kinds of differences, but the mere restriction of each stimulus to one kind of difference may produce so much similarity in other respects that this may become the more important fact. Furthermore, as we have just seen, even a fairly large difference in lightness may be masked by the presence of a diffuse boundary due either to the stimulus or to the vision of the observer, and the presence of large differences in other appearance characteristics such as texture, gloss, or other stimulus nonuniformities may make the color difference as such difficult to see. That contrast must thus, in one sense, remain quite a general concept does not mean that investigation of the variables that cause it is futile; it means only that there are many quite dissimilar causes of the phenomenon that cannot be lumped together without causing confusion. It seems likely that brilliance differences are an important cause of contrast, and recognition of this as a separate perceptual variable now needs careful investigation. It also seems likely, however, that the effect of eliminating brilliance differences to produce a similarity may turn out to be more important for aesthetic considerations than knowledge of its effect is to contrast perception. Brilliance is a very powerful factor, the relationship of which to the so-called simultaneous contrast effects is probably more important than the hue and lightness differences usually discussed. Thus when a spot is seen as gray against white and white against black this is a brilliance difference primarily, as is the enhancement of adjacent complementaries. Hue shifts, on the other hand, caused by noncomplementary pairs, probably are not and, in fact, may turn out to be maximized by a minimum brilliance difference. Incidentally, we might note that the word "complementary," used in this connection, really means that they do *not* cause a hue shift when placed next to each other rather than that they have some sort of relation to white.

MICRO- AND MACRO-STRUCTURE IN THE STIMULUS

Closely allied to the subject of contrast, but belonging ultimately to the subject of texture, is the perception of small or large inhomogeneities within the area of the stimulus itself. Almost any surface has a visible micro-structure and the fact that it is visible means that the stimulus is inhomogeneous. It is this micro-structure that has led to the unfortunate proposition that surface, volume, and aperture colors need to be treated separately. As we have seen, this is not true for color perception but is true for general appearance perception. As the size and contrast of the micro-elements increase, however, a number of factors come into play that need to be recognized. It is often stated (correctly) that as long as this micro-structure is not visible as such, what is seen is due to the simple additive mixture of the light from the different elements. This is often interpreted to mean that as soon as they are visible this ceases to be the case; this is often not true. Furthermore, when we think of micro-structure we are apt to visualize it in terms of elements of varying area in the same plane, whereas the effects that we see may be due to superimposed large areas each in itself seen as uniform, that is, the structure may be in depth rather than extension. While this may seem like a different concept rather than part of it, the two merge to such an extent that they should not be separated. Perhaps the best example is human skin. This has a micro-structure both in area and in depth that combines to give a unified perception that is almost impossible to imitate with an artificial stimulus; in fact this has been a major problem in plastic surgery. A truly high-gloss surface of either a solid object or liquid, or a liquid in a glossy container, are extreme examples. Lyrical statements about blue lakes often indicate a substitution of the reflection for the color of the water, which may be a dirty brown. In fact it is not usually possible to see the actual color of a body of water much below an angle of some 45° unless reflection is prevented. When the surface is roughened by waves and wind, this reflection is broken up and presents reflections from all parts of the sky as well as allowing some of the true water color to show. In less dramatic fashion a similar set of facts holds for all surfaces, whether or not they show visible roughness, and this is a true micro-structure as far as the stimulus is concerned. At the other extreme are the polished colored metals that, if clean, have no nonselective surface reflection and so present an essentially constant color stimulus, varying only in luminance if rough or curved. The practical impossibility of duplicating the color of polished gold with oil paints illustrates the extent to which this kind of micro-structure is integrated in perception.

As the size of the elements of area micro-structure is increased until the individual areas are clearly separate, we encounter a curious blend of the various perceptions. If the elements are distinctly different colors, there occurs at first the additive mixture of these colors such as would be obtained by spinning equivalent areas on a disk. As they become separately visible, however, both the edge-gradient effects and brilliance differences start to contribute to the effect, along with the colorimetric differences themselves. These effects are sometimes additive as far as the overall appearance is concerned and sometimes antagonistic, so that there are few general rules and essentially unlimited possibilities, as the enormous range of different woven fabrics shows. We can consider only a few tendencies but must keep in mind that the actual appearance of any one of them depends heavily on distance and the vision of the observer. Thus sharp boundaries around the elements will tend to increase brilliance, but this means much added gray for elements that are considerably darker than the average, and the Mach effect adds substantially to this. We perhaps see these effects in the very high purity colors found necessary by the painters of minatures of the last century, and we see them today in color photography in the greater colorfulness of large prints compared to small ones from the same negative. If the edges of the elements are diffuse, as in a woven material from "soft" yarns, both of these effects are decreased, the colors tend to blend rather than stand out, and we get a "soft" effect. On the other hand, a single sharp-edged strand running through the pattern, especially if lighter or of stronger color, may stand out as a major element of the pattern. Brilliance, as we have seen, depends heavily on hue, saturation, and lightness differences in relation to the sensitivity state of the eye. To be seen at all, however, the individual stimuli must be seen separately, so there has to be a continuous change in size of the individual areas from no perceived difference when they are so small that only additive mixture is seen up to a maximum when they are seen clearly and the eye is adjusted to their average. Such a change can be seen clearly by starting at a sufficient distance from any of the paintings of the "pointilliste" school and walking toward it. Even for quite clearly distinct patterns, however, the effects are not easily predictable. Thus I illustrated in my earlier book on color (Evans 1948), copying the illustration from von Bezold (1876), what I called the "spreading effect," in which a design in two colors shows black appearing as a component of the colors when they are separated by a black line and white when separated by a white one. This effect has the characteristics of additive mixture, as can be confirmed easily by simply drawing a blue line with a pen around any red figure on white paper, and it was for that reason that I spoke of it as "spreading." The fact that the effect occurs all the way across the pattern in a fairly sharp line where the black or white separators are not present, however, suggests that it is more than a little dependent on edge gradients of the Mach type. This whole subject of perception of colors in patterns would well reward a careful study; it has only been approached, and then only in piecemeal fashion. The justly famous investigations of Chevreul (1839) into the subject of simultaneous contrast, while they led to formulation of some of the principles involved (and to extravagant extrapolations of these principles that amuse us today), actually had little to do with the problem that he says led him to the study. That problem would seem to be more connected with additive mixture of the colors of intertwined silk threads.

Such intermingled fibers or disperse particles also involve the inter-reflections between colored fibers and particles that follow the laws of subtractive mixture, that is, successive selective absorptions. The whole subject is thus a very large one and so closely related to many of the objects of everyday life that lack of more precise knowledge is a serious gap in the subject of color perception.

DIRECTIONAL COLORS

One of the major causes of our ability to judge the distance and hence size and shape of nearby objects is the geometrical difference in the images in the two eyes caused by their separation. While it is debatable whether this fact is more important than the differences caused by head motion, it is certainly true that much of our perception of depth is due to changes in the geometrical relations within a scene that are evident when it is viewed from different directions. In discussions of the subject it is seldom pointed out, however, that changes in color also can play a considerable role in this process, especially as regards the shapes of the various objects. It is probably the exception rather than the rule when every point of any object is sending exactly the same amount of light to each eye although, of course, the difference is often negligible. Nevertheless, the fact that the shading of the light on such an object as a sphere is different for each eye must be the cause of the solidity seen in spite of the geometrical identity of the outlines. Such differences are usually due to the luminance changes of the individual points of the areas. There are many cases, however, in which there is also a difference in the spectral energy distribution. When this is produced by the directional characteristics of a surface, the perception produced is that of luster. That this is the cause of the appearance can be demonstrated easily by geometrically identical patterns viewed stereoscopically but with corresponding areas differently colored. Under these conditions even differences in luminance only will produce luster. It is apparent that luster is a considerably more general phenomenon than is usually supposed and, far from being restricted to highly specialized surfaces, plays an important role in form and depth perception.

That this role is not generally recognized in the literature is probably due to the fact that luster is thought of as an object appearance characteristic rather than as a color phenomenon. In fact, color nonhomogeneities in general, whether or not different to each eye, all represent borderline cases where the perceptions can be thought of as falling either within the restricted field of color as such or in the much larger general field of appearance. It is necessary in either case to distinguish between nonuniformities due to the separation of the eyes and those due to the stimulus alone. It is often implied that what is seen by the two eyes is integrated into a single response. While this is undoubtedly true of the mental concept formed from what is seen, it is not true of the responses to individual elements of the areas. The independence of the responses of the two eyes is a much overlooked and very important factor in visual perception. We note the subject here to distinguish it from differences caused by differing sensitivity states of the eyes.

METAMERIC ILLUMINANTS

It is assumed by all writers, as far as I know, that two light sources that are metameric produce the same eye sensitivity when used as illuminants. While there appears to be no reason to question the basis of the assumption, it is often not true in a real situation, because eye sensitivity is ultimately set by light reaching the eye from objects. We shall discuss the matter further under the heading of adaptation, but it may be worth noting here a few perceptual phenomena that are due to spectral energy differences of illuminants of the same color. We can restrict the discussion to those illuminants whose colors fall in the normal range of blackbody colors.

We can directly calculate the chromaticities and luminances of objects in a scene if we know the energy distribution of the source and the selective properties of the objects factors completely independent of the state of the observer.

As noted earlier, most of the artificial light sources available to early workers radiated because of incandescence and so had energy at all wavelengths, just as does daylight. Those that did not were considered curiosities, but the high luminous efficiency of mercury light led to attempts to make up for its spectral distribution deficiencies, and this led to the development of our present flourescent lamps. MacAdam pointed out many years ago (1938) that the maximum possible luminous efficiency for a colorless source would be obtained by concentrating all the energy in two narrow complementary bands at 490 nm and 608 nm. The efficiency of such a source compared to tungsten light at the same color temperature would be many times as great. Such a light source, however, if restricted to actual monochromatic lines, could produce only stimuli whose chromaticities lie along the straight line connecting them in the chromaticity diagram, unless there is actual fluorescence in the objects themselves. Thus for nonfluorescent objects all other stimuli on the diagram are eliminated. While the colors seen under such a source would have a somewhat greater range than this implies, because of adaptation and other effects, the range of possible colors would still be very seriously restricted; all pure nonfluorescent reds, for example, would be black. Mercury light has its energy concentrated predominantly in the blue and green regions so that maximum efficiency is produced by adding emission in the orange region. The somewhat artificial demand for higher light levels in large industrial areas, plus the cost of handling the heat load as radiation is increased, have thus forced us into the situation of having to accept light sources with inadequate energy distributions in terms of the range of stimuli they will produce from normal objects, in spite of the fact that the color of the illuminant itself is a satisfactory match for daylight. This is a deplorable situation in terms of working conditions for people and in terms of any task in which color plays an important role. It has led to a series of studies (see Wyszecki and Stiles 1967) on what is known as the "color-rendering index" of light sources by which they can be rated in respect to their distortion of "normal" colors. It is perhaps axiomatic that the only complete solution to the problem is a light source with continuous energy distribution and that all other solutions are compromises based on economics. The subject therefore lies somewhat outside our scope but it may be worthwhile to consider extreme cases in terms of the visual effects involved. We can restrict ourselves to two cases: in both, the energy is concentrated in two narrow bands, in the first case in the blue and yellow bands, in the second in the far red and cyan (blue-green) regions. Assume that they are metameric as sources and produce the same illuminance on the scene; a large sheet of white paper, say, would look the same in both and would match one illuminated by CIE Ill. C.

We shall consider later some of the effects these sources would produce (they are not all fully predictable as yet). I shall restrict myself here to luminance relations in the scene and the general hue-shift tendencies. Consider first the luminances of the two components in each source relative to each other. These luminances are given directly by the MacAdam moments. Thus if the two bands are monochromatic, the blue-yellow (B-Y) source will have perhaps 20 times the luminance in its yellow component compared to the blue, while the cyan-red (C-R) source will be more nearly balanced, with perhaps only 2 or 3 times the luminance in the red component compared to the cyan. Since most colorants tend to have higher reflectances at long wavelengths anyway, these facts will exaggerate these tendencies in the lightnesses of objects relative to each other, but the actual lightnesses will depend on their reflectances at the wavelengths involved. Thus, for example, most pure reds will be black under the B-Y source but very light under the C-R one. The importance of this lies not so much in the distortions of the normal lightnesses as in the brilliances that are produced. In a normal scene, illuminated by, say, Ill. C, even a red with a purity approaching 1 would normally have a luminous reflectance of less than 0.2 and sometimes very much less. Under the B-Y source such a color would approach 0 reflectance, limited only by its nonselective surface reflection. Under the C-R source, however, its luminous reflectance would approach something perhaps as high as 0.8. This can put such a color well up into the fluorent range! Similar considerations apply to all the selective surfaces involved; the general effect of such sources is not only a strong emphasis on the color due to their more luminous component but also a greatly extended brilliance range for the colors seen. Since yellow is so similar to white, this latter effect is not so startling with the B-Y illuminant; in fact the effect is very similar to that of sodium lights. The effect of the C-R source, however, has to be seen to be believed.

These effects decrease, of course, as the band widths of the components increase, but the tendencies remain noticeable all the way to continuous blackbody distribution and, to anticipate a little, are clearly in evidence, even after adaptation, in the relatively "cold" Ill. C compared to the "warm" Ill. A of unfiltered tungsten. They are very evident in the so-called "standard" high-efficiency fluorescent lighting of the B-Y type, particularly in the dark and washed-out appearance of reds.

Returning to our hypothetical sources with complementary monochromatic lines, we want to note an effect that occurs as soon as these line distributions are broadened into bands. When used as spectral components in the calculation of the chromaticity of the light source itself, these bands may be considered as each having a dominant wavelength, with the requirement that these be complementary. When used as sources of illumination, however, they can no longer be considered in this way; the important question becomes the wavelength *regions* in which there is energy. Consider a purely hypothetical case: suppose each band is widened so that it has a wavelength range of 10 nm and that objects can have any imagined selective reflectance. The chromaticities possible in the scene are now those included on the CIE diagram by a rectangle formed by connecting the *outer limits* of each band. This rectangle necessarily includes, and encloses, the Ill. C. Therefore, in theory, this light source can produce in the scene *all* dominant wavelengths, and this statement is true for all such light sources with finite band widths. The restriction on such a source as to what stimuli can be produced is thus not dominant wavelength but purity of the colors intermediate between bands; the wider the bands the higher the possible purities in these regions. This statement can be generalized to all light sources that are acceptably colorless but have discontinuous energy distributions. The discontinuities do not cause omission of possible dominant wavelengths; the requirement of course is the well known one that there must be three wavelengths present whose possible mixtures surround the white point.

Before leaving the subject of light sources with highly selective energy distributions, we might note a visual phenomenon that is particularly evident here but is true of all such stimuli. It is a basic assumption of colorimetry, and one which we have essentially implied throughout the book, that an isolated stimulus will have a unitary appearance regardless of the nature of its energy distribution. Expressed differently, this is the basic assumption that all metamers are indistinguishable. This is a necessary assumption for colorimetry and the fact that it is not strictly true does not lead to ambiguities in that subject. From the standpoint of perception, however, it is important to keep in mind that metamers *are* in general distinguishable from each other if their energy distributions are distinctly different, and that an energy distribution such as one of those above, where all the energy is concentrated in two narrow bands, is detectable as such without need of a comparison. This fact is due to the inhomogeneities of the eye as a receptor and to the fact that under normal conditions the eye is not kept in a fixed position relative to the stimulus. Mixtures of complementary wavelengths are often seen as separate colors superimposed, and the decision that such a mixture is neutral when seen as an isolated color is the decision that neither of them predominates. Many interior house painters are well aware of the difference between a "lively" and a "dead" neutral gray, and we do not need to elaborate the fact here; it is often overlooked, however, in discussions of applied colorimetry.

COLORED SHADOWS

One of the most striking phenomena in perceived color is the color of shadows cast by two illuminants of different chromaticity. This effect was known before Newton and was one of the basic facts of concern to Goethe. In its simplest form, if an opaque object is placed so that it throws separate shadows on a white surface illuminated by two lights of different colors, these shadows will always be seen in complementary colors—complementary in the sense that the hues are roughly opposite each other in any complete hue circle. At the same time (and this is not often mentioned in descriptions) the white surface itself is seen to be illuminated by the mixture of the two if there is a considerable difference between them; that is, in the general case, the two shadows, the illumination on the paper, and the two light sources themselves are all seen as of different colors. The phenomenon is not limited to shadows; the requirement is simply that there be two areas, each of which is illuminated by only one of the lights and that both of these are surrounded by the mixture color. A convenient way to study them is to project colored filters onto a screen by means of two projectors. If an opaque patch is placed on each filter so that they cover separate areas on the screen, the conditions are met; or alternatively the two patches can be on clear slides and the filters held over the projection lenses. By this means a very large number of color pairs may be examined quickly and the general nature of the effect studied. Since such projection can be done best in a darkened room, description of the actual stimuli involved is simplified. Without attempting anything but a rough description of what is seen, we can make a number of broad generalizations, at least one of which is rather surprising. In the first place, since the hues are the main point of interest, the effect is optimized if the luminances of (strictly illuminances from) the two sources are about the same. Under these conditions the apparent purities of the colors seen depend heavily on the purities of the sources themselves, but where there is a large discrepancy in the purities of the sources, it is the purity of the purer one which governs the effect. Thus if, as isolated colors, one source is red and the other achromatic, it is the purity of the red that determines the apparent purity of *both* the red and cyan patches seen. [I use the term "apparent purity" here, not because it is properly descriptive, but to call attention to the fact that it is not known whether the effect is a saturation or a brilliance change or both. It would be an important contribution to establish this point—presumably by inter-eye studies.] Furthermore, it is found that the colors of the two sources do not have to be distinctly different to produce the "complementary" hues. Two sources of slightly different color temperatures in the range normally accepted as colorless will produce "blue" and "yellow" patches on a "white" ground; and this fact provides a fairly sensitive test for such a difference. Two purities of the same dominant wavelength will be seen in the surround as the intermediate purity of the mixture; one patch will be roughly this same hue at higher effective purity and, again, the other will be of "complementary" hue.

As far as hue is concerned, these effects can all be predicted by a single generalization as startling in its implications as in its predictions: if the chromaticities of the two sources are plotted on the CIE diagram and the line connecting them extended to the spectrum locus in both directions, the intercepts of this line indicate the wavelengths (or complementary wavelengths) that, seen as *isolated* colors, would be of roughly the same hues as the two patches, regardless of where the source points lie on the diagram. Since the mixture chromaticity of the two sources lies between them on this same line, this leads to a definition of the complementary with respect to this mixture point. But note that this is loose terminology; the mixture is not seen as achromatic and wavelength should not be used to describe hue. Nevertheless it is a valid rough description of what is seen. For many pairs of sources the hues seen are far from complementary in the ordinary sense. We shall return to these effects presently but should note here two other points. If the luminances of the two sources are considerably different, the effects are not easily discernible and appear to be somewhat unstable; this area needs study. When the sources are of comparable luminance, however, the effect is instantaneous as far as the observer is concerned and completely stable with respect to eye movement. It does not matter, for example, whether one source is turned on first and then the other or both are turned on together; the colors appear immediately when both illuminants are present, and the colors are seen as uniform over the areas of the spots, with the background also uniform in the neighborhood of the spots.

We should note here that colored shadows are only a particularly spectacular case of a general phenomenon which is usually given the name "simultaneous contrast." If only the actual stimuli corresponding to the surround and *one* of the spots are present, the same color is seen in the spot as when the other is present. In the general case of a single spot, all possible hues can be produced in a normally achromatic stimulus by changing the chromaticity of the surround only, the spot always appearing "complementary" to the hue of the surround. The effect is just as immediate and just as stable, except that by fixating the spot for an appreciable length of time some changes may be produced, as is true of any pattern.

The phenomenon of colored shadows and the more general phenomenon of simultaneous contrast lead to the famous generalization first stated by Chevreul (see Chevreul 1839.). As translated in 1854, he wrote: "All the phenomena I have observed seem to me to depend upon a very simple law, which, taken in its most general signification, may be expressed in these terms: In the case where the eye sees at the same time two contiguous colors, they will appear as dissimilar as possible, both in their optical composition and in the height of their tone." We have seen that this can now be extended to all five color perceptions.

AFTERIMAGES

We noted in an earlier chapter the changes in the sensitivity of local areas of the eye brought about by continued viewing of a stimulus. The extent of the change produced and the duration of the change after exposure are both heavily dependent on many factors, and there is so little quantitative information on the subject that detailed discussion from the standpoint of perception is not worthwhile. We do need to consider it, however, in terms of the colored shadow phenomenon we have just discussed and the subject of adaptation we shall discuss next.

Colored shadows and afterimages are quite different kinds of phenomena perceptually, but the fact that the word "complementary" is used loosely to describe both effects tends to suggest a common cause, that is, a change in receptor sensitivities. It seems likely that this is not true, at least in any simple way. The production of an afterimage is very obviously a time-dependent phenomenon, both in the initial formation and in the decay of the effect; it is favored by relaxed viewing and a fixed point of view and may be renewed by looking back at the stimulus. It is also easily demonstrated as a local retinal effect by the fact that it covers a fixed angle of view in the external field. Colored shadows have none of these characteristics; they are instant, static, and independent of the motion of the eye. The two effects, however, do often occur simultaneously. If a patch of color on a larger ground is viewed steadily for an appreciable length of time, involuntary movements of the eye will produce the afterimage of the patch along its edges, always accompanied by a dulling of the appearances of both the patch and the surround. If both phenomena are to be considered changes in the sensitivity state of the receptors, then, at the very least, they have to be postulated as different *kinds* of changes.

It seems to me that a very real and perhaps large contribution to the whole subject of color perception could be made by a thoroughgoing quantitative comparison, necessarily by inter-eye studies, of the similarities and differences of the color changes produced by these two effects. It is customary to refer to both as "adaptation" effects, and it is this, perhaps, that has suggested that the same mechanism underlies both. It seems likely that adaptation of the type exemplified by the von Kries equations would, in fact, be found to quantitatively explain and predict afterimages under controlled time conditions. It appears to have been adequately demonstrated that to the extent that colored shadows are representative of more complex viewing situations, they cannot be predicted by this approach.

ADAPTATION

This brings us to the general subject of adaptation. If the meaning of the word is restricted to those cases in which there is a demonstrable change in the sensitivity of the eye caused by the presence or absence of a physical stimulus, then quite definite statements can be made about it, most of which have been touched on earlier. Unfortunately, the word is often used in cases where "become accustomed to" or "accept" would be better terminology. There is thus a tendency to fall back on it as an "explanation" of anything for which the cause is not immediately apparent. In my previous book on color (Evans 1948), following the literature, I divided adaptation effects into local, lateral, and general, represented roughly by afterimages, simultaneous contrast, and adjustment to radically different overall illuminations. The implication was that all effects of these types are caused by physiological or neurological processes taking place in the receptor mechanism and hence presumably derivable from our very considerable knowledge of the characteristics of these processes. It is an assumption, not a necessary fact, that all perceptions of this type are due solely to processes at the receptor levels. It is for this reason, as I implied earlier, that one of the problems in relating color perception to colorimetry is the comparative naivete of our concept of adaptation. The fact that a phenomenon is repeatable and acceptably alike for normal observers does not mean that it is necessarily caused by neural responses or interactions. The perception of any difference between two stimuli, when appreciably greater than just perceptible, involves a type of perception different from recognition of the response to a single stimulus. While it cannot be argued that neural processes are not involved, neither does it follow that rules followed by one are applicable directly to the other. Thus the angle subtended at the eye by an isolated spot is directly observable; but when this spot is seen in a context that includes real objects, then the *size* of the spot is immediately related to its apparent distance from the observer and its angular subtense becomes immaterial and, in fact, very difficult to judge. I am suggesting not that perceived hue is a judgment but that hue may turn out to be perceived as a difference phenomenon rather than as a direct response to the stimuli in some cases, particularly those in which adjacent colored areas are involved. Specifically, I am suggesting that the perceptions of hue in complex situations may well be with respect to some state set up at fairly high levels in the brain by the total situation. This is all quite vague but I introduce it deliberately to emphasize the fact that we do not have enough quantitative knowledge of the causes of hue perception to distinguish among the possibilities. Without more specific information we are simply making guesses, based on probably fallacious assumptions. I repeat that such firm knowledge can probably be obtained only by carefully controlled inter-eye comparisons, taking full account of all five perceptual variables.

For similar reasons I feel it is futile at this point to attempt an analysis of the very considerable amount of information that has already been published based on this technique. None of it, to my knowledge, has dealt with perceived hue as such; it has all been based on inter-eye metamerism under various conditions; and while this operation is the necessary technique, the studies have not been directed toward the cause of the perception so much as toward trying to apply existing knowledge to predict them. What is needed is new knowledge. What are all the conditions that will produce adaptive metamers, that is, the same perceived color in a single area in the field of view? Or, less ambitiously, what are all the relationships between two stimuli that will result in the same perceived color in a central spot surrounded by a much larger area? While we know that each stimulus may be substituted by any of its metamers without changing what is seen, we do not *know* (it has not been proven) whether the result is unique, that is, whether the identical perception of the central spot with respect to *all* color variables can be obtained by only one particular pair of perceived colors. We confidently assume that there are an infinite number of such combinations, and there are almost certainly at least two, as we discussed earlier.

These remarks apply primarily to lateral adaptation effects and suggest that they may be different in kind from local and general effects. This possibility and our lack of real knowledge except in highly specialized aspects of it make discussion of general adaptation necessarily vague, but we can note certain aspects that are well established. In the first place, it is quite apparent from common experience with daylight that general adaptation to the luminance level of a scene does not, in itself, cause hue shifts in the stimuli, and hence, in this sense, we can say that brightness and chromatic adaptation are independent. On the other hand, it is equally evident that color discrimination and the total range of colors seen is far greater at the upper photopic levels than at the lower so that it is an oversimplification to say that the two are independent of each other. Much of the descriptive information we have in this field is uninterpretable because of the use of the word "saturation" to apply to both saturation and brilliance, as we have defined the terms. Until the two have been separated by definitive subjective studies, we cannot make statements on the effect of absolute levels, and similar remarks apply, as we have seen, to chromatic adaptation. We can only conclude that the next great advance in colorimetry, particularly in the field of perceived color differences, will come about through a more complete understanding of the true nature of the adaptation process and that in studies leading to this knowledge, separation of the perceptions of saturation and brilliance will necessarily be a key factor.

In this respect we might note another factor, representing really an historical point of view rather than a misunderstanding, that has tended to obscure general observations in the field. It was customary for many years to describe adaptation effects, particularly local ones, as "fatigue" of the eye. While few modern writers now use the term, the concept has tended to persist; adaptation tends to be considered always a *loss* of sensitivity. Many adaptation phenomena are far more profitably considered a gain in sensitivity. Whatever concept is adopted as to the physiological nature of adaptation, the basic fact would appear to be that the sensitivity of the eye always tends to rise or fall until it is in equilibrium with the existing stimulus. This implies, of course, that the sensitivity of the eye is greatest at low illumination levels and less at higher ones and this is certainly true on an absolute basis. The statement, however, gives a misleading impression of what is seen. In terms of photopic levels only, as the luminance level of the stimuli in a scene increases there is a downward movement in overall sensitivity but a steeply increasing *range* of perceptions produced by the stimuli. It seems that the important fact is the distance of the stimuli above the chromatic threshold. Expressed somewhat differently, it seems that the equilibrium between the force tending to increase eye sensitivity and the force of the stimulus holding it back is as much a local phenomenon as a general one so that at higher general levels the sensitivity of the eye to low reflectances is greater, relative to, say, white, than it is at lower levels. This concept also applies to the color perception variables produced by a single stimulus within the scene. Although the matter is confused by all manner of changes in the actual physical stimuli, it is common experience that far more colors are distinguishable in bright light than dim even when both are well within the photopic range. In addition to this characteristic, general adaptation must also be considered as more dependent on stimuli whose images fall near the fovea and as changing rapidly in both sensitivity directions as the eye is moved about over the scene, sensitivity dropping more rapidly in bright areas and rising somewhat more slowly in darker areas. These changes take place in ordinary scenes within fractions of a second, the time probably less at high general levels than at low. In a sense, it is as though the force tending to raise sensitivity is greater the greater the force holding it back.

It is with respect to chromatic stimuli that this concept of a force tending to increase sensitivity is most descriptive, because the force can be considered spectrally selective. Consider again the viewing of the green spot on a white surround to produce an afterimage. Fixation on the spot holds the image steady on the retina, and over this area there is stimulation primarily from the green part of the spectrum. Since there is relatively little stimulation in the blue and red, the sensitivity level can rise *higher* before equilibrium in these regions; this "explanation" accords well with what is seen when the afterimage is directed toward yellow and cyan second areas. A still better example in which this point of view is helpful occurs in what I have called the Helson-Judd effect (Evans 1948). In this effect a whole scene is typically illuminated by a chromatic source of high purity and midrange photopic intensity. Under these conditions, after adaptation, shadow and reflectance differences tend to disappear and are replaced by hue differences, the lighter areas of the same hue as the illuminant and the shadows "complementary" but of the same brightness. It is apparent that we have here not only a rise in sensitivity of the eye to the complementary part of the spectrum but probably also some "cross talk" either at the retina or at higher levels. Here again it seems as though carefully controlled studies of the actual hues in such effects would go far toward defining receptor sensitivities.

The basic, unresolved question with respect to adaptation can be stated as follows: "For the case of two stimuli, one of which is of much greater area and surrounds the other, can the values of the five perceptual variables produced by the central stimulus be predicted from present knowledge by an equation involving the six colorimetric variables of the two stimuli? " We have seen that at least hue and brilliance seem to require the introduction of a function that is, at present, known only incompletely and that does not appear derivable from the existing established data of colorimetry. On the other hand, there is no evidence that the result is capricious and so no reason to believe that such a mathematical relationship cannot be established. I think we must conclude that we cannot state such a relationship today and that the reason lies in our lack of understanding of the phenomenon called adaptation, perhaps more with respect to its *nature* than its quantitative relations, but perhaps also in our assumption that there is a *single* mechanism involved.

THE LAND EFFECTS

Our lack of knowledge in this field has been most brilliantly called to our attention in recent years by the remarkable demonstrations of E. H. Land (1959). In these he has used as stimuli two superimposed projected pictures, identical except for their relative luminances in each area, each projected by lights of different chromaticity, with the picture usually representing real objects. Since these are projected in a semidarkened room they are the main stimuli visible and so control the adaptation of the eye to a large extent, the whole area acting as an isolated stimulus, the various parts as related stimuli, and the whole usually meaningful with respect not only to real objects but also to illumination. His arrangement of stimuli is perhaps the most complex, from the standpoint of perception, that has ever been investigated seriously. It is more complex in one sense than a real scene, illuminated by his two sources, would be because the relative luminances of the individual areas are determined not only by their reflectances to the sources but also by the light source, filter, and film sensitivities used in making his transparencies and by the contrasts to which they are processed. These facts, however, are largely irrelevant to what is seen; they act only to modify the effective spectral reflectances of the original objects, reducing them all to a function of their red-green reflectance ratios, both red and green being quite broad bands. On projection this relationship is always retained, the longer effective wavelength source always projecting the "red" record.

We can add little to the very considerable literature that has developed around these demonstrations. For contemporary analyses the reader is referred to the articles by Judd (1960), Walls (1960), and Sheppard (1968). It is apparent that most of the color perception phenomena that occur in a real scene can be evoked by such projections, and it follows that various ramifications of the projections will emphasize one or more of all of the factors we have considered. However, since we have concluded that we cannot yet predict the simple case of one stimulus surrounded by another there is little hope of a *complete* explanation based on present knowledge.

We can, however, look at certain of the demonstrations for the light they throw on some aspects of color perception. They add a number of facts that any overall theory will eventually have to encompass. The real problem is to sort out what is actually new and what is only apparently new because of the novelty (with respect to the literature) of the picture approach. We might note, incidentally, that the fact that many of the demonstrations are of recognizable real objects probably contributes very little to the actual colors seen but greatly to the surprise of the viewer. We have already considered the ability of the observer to compensate for the viewing situation and his ultimate analysis of object colors primarily in terms of the unique hues. This is all in full play in the demonstrations but does not prevent conscious observation of the actual colors seen, *particularly* in a picture, as opposed to real objects.

Perhaps the most striking fact of the demonstrations is the necessity that *both* separation images be projected. If either record is projected alone, by either source, with nonimage light from the other source covering the screen, neutral objects are seen. These may be thought of as illuminated by colored light, as seen through a colored filter, and so on, depending on circumstances, but no colors other than those predictable from the mixture colors of the two sources are seen. [Perhaps I should note here that while I have studied many of the "Land effects" in my own laboratory and have encountered many of them in my own work, what can only be described as a series of circumstances amounting to fate has prevented my seeing any of his demonstrations personally.] If the two sources are, say, red and white, these mixture colors represent a purity series from 0 to that of the red source (neglecting stray light from the room). We have already seen that in the case of colored shadows (or spots) this is not what would be seen. The explanation of this discrepancy necessarily lies in the fact that the observer sees the situation as such. As a matter of fact the statement is equally true of what an observer sees looking at a real scene through a high purity red filter; he sees nothing but neutral objects and a red hue. This still holds if white light reflects from the front surface of the filter. Yet as sources for colored shadows, red, white, pink, and a cyan of almost equal strength to the red are seen. In a way the situation is almost exactly analogous to that of the perception of colored illumination as separate from that of the selective characteristics of objects in a natural scene. When we attribute the color to the illumination and there is no inconsistency, we see only the one color; when we can attribute it to the objects only, we see relative hues, and it may well be that this is, in fact, the demonstration we are seeking, that is, that hues seen in a related situation are just that, the perception of *relative* effective wavelength. In the "lightness constancy" situation we see local relative luminance and surface reflectance quite independently from their relation to the scene as a whole. In the Land situation we see relative local wavelength relations as hues, *when we can attribute them to objects* (or stimuli in the case of colored shadows).

This, however, is an extreme psychological point of view. If we consider the stimuli more carefully we find there is an obvious, fundamental difference when the two images are present. While it is true that in theory at least the same *ratios* of the two stimuli can be present in both cases and that these ratios are seen as different hues in one case and different lightnesses in the other, the statement is misleading. Consider first the case of red-white projection with the image in the red beam only. Knowing the luminances produced on the screen by the two sources and the transmittance of each point on the image slide, we can write a fraction for the ratio of red to white luminance at each point on the screen. These fractions all have the same denominator because white is constant throughout. The value of the fraction will reach 1 only if the projectors are matched and a fully lighted white in the scene is represented by 0 density in the slide (or if the projectors are adjusted for this area). The fraction can be made to exceed 1 by dropping the intensity of the white, and this does not change the fact that only one hue is seen. The denominator, however, remains constant for all conditions. It follows that what is seen is not determined primarily by this fraction but by the fraction of the red component relative to *its* maximum; that is, the red and white components are seen separately as far as an image is concerned; the image, as such, is wholly in the red. Consider now the case where both images are present and, for simplicity, consider that any neutral area in the scene is represented by equal luminances of the two beams on the screen. The image now resides in both beams and they are indistinguishable for form and area. If we now write fractions in terms of the actual luminances of the two beams for each area, the denominator of each fraction will, in general, be different for each area. If these fractions are reduced to decimals they may well have the same values as in the previous case, but they do not represent the same physical facts. In this case they represent that fraction of the *image* due to the red as compared to the white, and this *is* what is seen in this case.

If, for simplicity of description only, we make the assumption that the eye is adapted to an equal part mixture of the two beams and sees this as neutral, that is, the mixture has become the reference "white" for the system just as it was in the original scene, then fractions higher than 1 will represent areas with preponderant red effective reflectances in the scene and those below 1, preponderant green. Since the sources themselves, seen as colored shadows or spots, would be seen as red and cyan, the two images thus generate the whole range possible with these colors. Similar considerations hold for all pairs of projection chromaticities.

Note that a *necessary* requirement for this perception is that the two projected pictures be seen as a *single* image. As Land pointed out, and as is well known to anyone who has worked with additive color photography, if the two images are not "in register," that is, obviously do not coincide, then the perception immediately reverts to the perception of two images each in its own color. The extreme case of this, which he also reports, consists in having the two images represent different scenes. The requirement is that the observer perceive what is seen *as a mixture.* Like illumination perception, it is strange only if one takes an extreme one-to-one position with respect to stimulus-perception relations.

The question thus becomes, "Why are hues that are not associated with mixtures of the two projection colors also produced?" This obviously involves the nature of the adaptation state and cannot be answered quantitatively at the present time. There are some leads along these lines, both from color photography and from other Land experiments, and we can review them briefly. We note first that hue is a qualitative perception and that to assign a hue name requires only that a perception be recognizable as such. Colorimetrically the difference between two colors seen as very different in hue may be quite small; it is the direction of the colors from the stimulus called achromatic that matters.

The fact that more hues than were suggested by colorimetric analysis of the stimuli were produced by "two color" color photography was well known to workers in the field long before the Land demonstrations. I remember that L. T. Troland was much concerned about the fact that he could not explain the colors he saw in some of the Technicolor pictures produced by the then two color subtractive process being used. As production manager and scientific advisor for Technicolor, as well as professor of psychology at Harvard and chairman of the Colorimetry Committee of the Optical Society of America, he was in a good position to realize the magnitude of the discrepancies. It was also well known at the time that, for the best results, the subtractive primaries of such a process should not be "complementary," that is, no combination should produce an actual neutral in the film although apparent neutrals would be seen on the screen.

These facts and others were ascribed to adaptation processes and lumped with the color shadow phenomenon, but with general acceptance of the fact that they remained to be explained. Thus, when in 1942 I included in a lecture called "Visual Processes and Color Photography" a demonstration of the colors (which included blue and purple) produced by additive projection of two color pictures through yellow and red filters, I made so little point of it (three color subtractive then being well established) that it is hardly mentioned in the subsequent published form (Evans 1943) of the lecture. We need now to consider this effect, briefly, from the perceptual point of view, and we need consider only one aspect of it. Both this demonstration and the many, more elegant ones of Land were produced by sources that can be considered not only monochromatic but lying on a portion of the spectrum locus in which their mixtures also have a purity of 1 and lying along the yellow-orange-red part of the spectrum. The hues seen, however, include greenish, bluish, and purplish colors. Even though we know that as colored shadows the yellow will look green and the red very red, we are not prepared for the presence of colors having blue as a component. We also know that if the same wavelength is compared at two luminances, there will be a very appreciable hue difference between them, the direction depending on the wavelength, but these changes are along the spectrum locus and can be compensated for by a wavelength change. We can also rule out room light as a factor because the effects occur in complete ambient darkness as well as in room light. What cannot be ruled out is what I have called the Helson-Judd effect, that of the spectacular rise in sensitivity of the eye to the "complementary" of a very pure adapting stimulus, and in these cases this is blue sensitivity. We can thus postulate blue response for lower luminance due to this phenomenon; and, at the moment, this may be as close as we can get to a physiological explanation. In any case, I think we have to conclude that the concept of hue as a direct perceptual response to a stimulus is either inadequate or incorrect. Hue is, perhaps, rather the direct perception of the *direction* of departure of the stimulus from some reference stimulus. Such a concept appears to fit all situations, including that of isolated colors where, as we have seen, the reference is actually the equilibrium state against a 0 stimulus.

We may conclude this note with mention of a further extension of this demonstration by Land, one that Sheppard makes considerable point of, in which the two primaries are monochromatic "yellows" at 570 and 580 nm. The demonstration is surprising only in the small wavelength difference of the primaries. (I think I am correct that Land also produced effects with the two lines of the sodium doublet.) Granted the other effects these would seem to follow. They call attention, however, to what Sheppard calls the "yellow anomaly." We noted earlier that this anomaly consists largely in the fact that yellow is so much like what we call white. It calls attention, however, to a fact that cannot fail to impress anyone seriously concerned with color perception, namely, that it is blue that is radically different from the other color perceptions. While I have adhered as strictly as I could to my resolve to not mention color perception theories, I note here that it is certainly easy to accept the underlying facts subsumed by the early Ladd-Franklin theory (1929), whatever its phylogenetic merits may be. In this theory yellow and blue are thought of as separate responses and green and red as differentiations within the yellow.

I am reminded of an occurrence of many years ago which may not be too out of place to relate here. After a conference on vision at Ohio State University, Selig Hecht and I had luncheon together and spent some hours waiting for our train. To pass the time I brought up the fact that it was easy to imagine three color responses from two receptors (the cones and rods) if one made a few arbitrary assumptions. To assume that rods respond according to the luminous efficiency curve, that cones respond á la Hering to red or green, and that blue is the perceived difference in excitation of the rods and cones would appear to be all that was necessary. We had a very pleasant few hours batting the idea around and I never saw Hecht again. It turned out, however, that he did not forget the occasion; some years later he wrote a review of a book that was based on a somewhat similar proposal. In his customary caustic fashion, he said that while such speculations may prompt very interesting dinner conversations they had no place in a published book. So be it!

THE LANGUAGE OF DRAWING AND PAINTING

Perhaps the most analytical statement in the literature of color on the inadequacy of three variables to encompass the facts of color perception is found in the book with the above title written by Prof. Arthur Pope (1949) of the Fogg Art Museum of Harvard. In many years of teaching color to students of art, and in two successive books, he struggled to reconcile the terminology which he found necessary to describe the work of the great artists with the "facts" as set forth by scientists and psychologists. Accepting, as he did, the belief in three variables, he was totally unable to get into words the inadequacy of all proposed schemes of arrangements of colors.

His book is eloquent testimony to both this inadequacy and the utter turmoil of terminology that has resulted, over the years, as the subject has been approached from one direction or another by different workers, each seizing on a different set of three variables they believed to be sufficient.

It is in this sense that we should read, for example, Pope's statement: "In order to define any tone accurately from the visual or psychological point of view, it is only necessary to state its hue, value and intensity (chroma); but the two further factors, which I shall call purity and brilliance, must be considered if one is to possess a complete understanding of the subject."

It is unfortunate that Pope's own terminology, made necessary by his valiant attempt to squeeze what he knew to be the facts into three dimensions, precludes any simple translation of his statements into our variables. There is no question, however, of the fundamental soundness of his ideas, nor of the fact that a complete rewriting of that portion of his book in terms of the four variables, hue, saturation, brilliance, and lightness as we have developed them, would remove most, if not all, of the ambiguities he encountered. Carrying out such a work would be a remarkable contribution to the understanding of the arts; it is regrettable that it is so far beyond our present scope.

It is interesting and noteworthy that the two clearest indications of the inadequacy of three variables to describe perceived color came from such diverse approaches. Thus, it was in his attempts to space all possible object colors uniformly that Judd found that three variables would not work, and he suggested that the solution involved (in effect) the addition of a fourth dimension by the use of a background "halfway between" any two. Pope arrived at it by way of the terminology that had been used over the years and by his inability to resolve that terminology into three adequate words.

Pope appears to have come near to a flat statement that four terms are necessary (both he and Judd implicitly assumed that brightness, our fifth, was irrelevant, as it is) in his bibliography, in the note that follows the listing of my book, *An Introduction to Color.* Concerning this he says: "It may be noted that the matter of purity (as defined herein) is completely ignored." The "herein" refers to his own definition of "purity" which is closely akin to, if not identical with, our "brilliance." This is not only a statement that a fourth variable is needed but also a completely valid and penetrating criticism of my book. It is my hope that this present work sufficiently remedies the deficiency.

BIBLIOGRAPHY

Adams, E. Q.

1923. A theory of color vision. *Psych. Rev.* **30,** 56.
1942. X-Z planes in the 1931 ICI system of colorimetry. *J. Opt. Soc.* **32,** 168.

Beck, Jacob

1972. *Surface Color Perception.* Cornell University Press, Ithaca, N. Y.

von Bezold, W.

1876. *The Theory of Color.* Prang, Boston.

Billmeyer, F. W., Jr., and **Max Saltzman**

1966. *Principles of Color Technology.* Interscience, New York.

Birren, Faber

1965. *History of Color in Painting.* Van Nostrand Reinhold, New York.
1967. *See* Chevreul, **1839.**

Boring, Edwin G.

1942. *Sensation and Perception in the History of Experimental Psychology.* Appleton-Century-Crofts, New York.

Burnham, R. W., R. W. Hanes and **C. J. Bartleson**

1963. *Color: A Guide to the Basic Facts and Concepts.* Wiley, New York.

Chevreul, M. E.

1839. *The Principles of Harmony and Contrast of Colors.* (Introduction and Notes by Faber Birren.) Van Nostrand Reinhold, New York (**1967**).

Cornish, Vaughn

1935. *Scenery and the Sense of Sight.* Cambridge University Press, Cambridge.

Evans, R. M.

1943. Visual processes and color photography. *J. Opt. Soc.* **33,** 579.
1948. *An Introduction to Color.* Wiley, New York.
1949. On some aspects of white, gray and black. *J. Opt. Soc.* **39,** 774.
1954. The expressiveness of color. *Ed. Theater J.* **6**(4), 327.
1959. *Eye, Film and Camera in Color Photography.* Wiley, New York.
1959. Fluorescence and gray content of surface colors. *J. Opt. Soc.* **49,** 1049.
1964. Variables of perceived color. *J. Opt. Soc.* **54,** 1467.
1967. Luminance and induced colors from adaptation to 100 millilambert monochromatic light. (Letter to Editor). *J. Opt. Soc.* **57,** 279.

Evans, R. M. and **J. Klute**

1944. Brightness constancy in photographic reproductions. *J. Opt. Soc.* **34,** 533.

Evans, R. M. and **B. K. Swenholt**

1967. Chromatic strength of colors. I. Dominant wavelength and purity. *J. Opt. Soc.* **57,** 1319.
1968. Chromatic strength of colors. II. The Munsell system. *J. Opt. Soc.* **58,** 580.
1969. Chromatic strength of colors. III. Chromatic surrounds and discussion. *J. Opt. Soc.* **59,** 628.

Faulkner, Waldron

1972. *Architecture and Color.* Wiley, New York.

Goethe, J. W.

1840. *Theory of Colours.* (translated by Charles Eastlake.) Massachusetts Institute of Technology Press, Cambridge, Mass. (**1970**).

Gombrich, E. H.

1960. *Art and Illusion: A study of the Psychology of Pictorial Presentation.* Phaidon, London.

Graham, Clarence H. (ed.)

1965. *Vision and Visual Perception.* Wiley, New York.

von Helmholtz, H.

Treatise on Physiological Optics, 3rd ed. Translated by J. P. C. Southall. Dover, New York (**1962**).

Helson, Harry

1964. *Adaptation Level Theory.* Harper & Row, New York.

Hering, Ewald

1905–1920. *Outlines of a Theory of the Light Sense.* Collected and translated by Leo M. Hurvich and Dorothea Jameson. Harvard University Press, Cambridge, Mass.

Hesselgren, Sven

1954. *Subjective Colour Standandization,* Almqvist & Wiskell, Stockholm.
1955. Description of a colour atlas based on the phenomenological analysis. *Congres F.A.T.I.P.E.C.* (May, 1955) 171–174.

Hurvich, Leo M. and **Dorothea Jameson**

1951. A psychophysical study of white. *J. Opt. Soc.* **41,** 487.
1953. Spectral sensitivity of the fovea. I. Neutral adaptation. *J. Opt. Soc.* **43,** 485.
1953. Spectral sensitivity of the fovea. II. Dependence on chromatic adaptation. *J. Opt. Soc.* **43,** 552.
1954. Spectral sensitivity of the fovea. III. Heterochromatic brightness and chromatic adaptation. *J. Opt. Soc.* **44,** 213.
1955. Some quantitative aspects of an opponent-colors theory. I. Chromatic responses and spectral saturation. *J. Opt. Soc.* **45,** 546.
1955. Some quantitative aspects of an opponent-colors theory. II. Brightness, saturation and hue in normal and dichromatic vision. *J. Opt. Soc.* **45,** 602.
1956. Some quantitative aspects of an opponent-colors theory. III. Changes in brightness, saturation and hue with chromatic adaptation. *J. Opt. Soc.* **46,** 405.
1964. *See* Hering, **1905–1920.**
1966. *The Perception of Brightness and Darkness.* Allyn and Brown, Boston.

ISCC–NBS

1955. *Method of Designating Colors and a Dictionary of Color Names and Supplement.* ISCC–NBS Color-Name Charts of the Centroid Colors. National Bureau of Standards Circular 553 and supplement.

Judd, D. B. and **K. L. Kelly**

1939. Method of designating colors. *J. Res. Natl. Bur. Stand.* **23,** 359.

Judd, D. B.

1951. Basic correlates of the visual stimulus. In *Handbook of Experimental Psychology* (S. S. Stevens, ed.). Wiley, New York.
1952. *Color in Business, Science and Industry.* Wiley, New York.
1960. Appraisal of Land's work on the two-primary color projections. *J. Opt. Soc.* **50,** 254.
1969. Ideal color system. *Palette* (*Sandoz*), nos. 29, 30, 31. Sandoz Ltd., Switzerland.

Judd, D. B. and Gunter Wyszecki

1963. *Color in Business, Science and Industry,* (2nd Ed.). Wiley, New York.

Katz, D.

1911. *The World of Color.* Kegan, Paul, Trench, Trubner, London **(1935).**

Kelly, K. L.

1943. Color designations for lights. *J. Opt. Soc.* **33,** 627.

Ladd-Franklin, C.

1929. *Colour and Colour Theories.* Harcourt Brace Jovanovich, New York.

Land, E. H.

1959. Color vision and the natural image, Parts I & II. *Proc. Natl. Acad. Sci.* **45,** 115 and 636.

1964. The retinex. *Am. Sci.* **52,** 247.

Le Grand, Yves

1957. *Light, Colour and Vision.* Wiley, New York.

MacAdam, D. L.

1935. Maximum visual efficiency of colored materials. *J. Opt. Soc.* **25,** 361.

1938. Photometric relations between complementary colors. *J. Opt. Soc.* **28,** 103.

1942. Visual sensitivities to color differences in daylight. *J. Opt. Soc.* **32,** 247.

1956. Chromatic adaptation. *J. Opt. Soc.* **46,** 500.

1970. *Sources of Color Science,* Massachusetts Institute of Technology Press, Cambridge, Mass.

Mach, E.

1886. *The Analysis of Sensations.* Translated by S. Waterlow. Dover, New York (**1959**).

Mackinney, G. and **A. Little**

1962. *Color of Foods.* Avi, Westport, Conn.

Newhall, S. M.

1940. Preliminary report of the OSA subcommittee on the spacing of the Munsell colors. *J. Opt. Soc.* **30,** 617.

Nickerson, D.

1936. The specification of color tolerances. *Text. Res.* **6,** 509.

Nickerson, D. and **W. Granville**

1940. Hue sensibility to dominant wavelength change and the relation between saturation and colorimetric purity. *J. Opt. Soc.* **30,** 159.

OSA Committee on Colorimetry

1953. *The Science of Color.* Crowell, New York.

Ostwald, W.

1931. *Color Science.* Translated by J. S. Taylor. Windsor & Newton, London.

Parsons, J. H.

1924. *An Introduction to the Study of Colour Vision,* 2nd ed., Cambridge University Press, Cambridge.

Pope, A.

1949. *The Language of Drawing and Painting.* Harvard University Press, Cambridge, Mass.

Priest, I. G.

1926. Blue sky and white snow. *J. Opt. Soc.* **13,** 308.

Ratliff, F.

1965. *Mach Bands: Quantitative Studies on Neural Networks in the Retina.* Holden-Day, San Francisco.

Ruskin, John

1857. *The Elements of Drawing.* Dover, New York (**1971**).

Sheppard, J. J., Jr.

1968. *Human Color Perception: A Critical Study of the Experimental Foundation.* Elsevier, New York.

Sinden, R. H.

1923. Studies based on the spectral complementaries. *J. Opt. Soc.* **7,** 1123.

Stevens, S. S. (ed.)

1951. *Handbook of Experimental Psychology.* Wiley, New York.

Troland, L. T.

1929. *The Principles of Psychophysiology.* Van Nostrand Reinhold, New York.

Walls, G. L.

1942. The vertebrate eye. *Cranbrook Inst. Sci. Bull.* No. 19.
1954. The filling-in process. *Am J. Optom.* **31,** 329.
1960. Land, land. *Psych. Bull.* **57,** 29.

Woodworth, R. S. and **H. Schlosberg**

1954. *Experimental Psychology,* 2nd ed. Holt, Rinehart & Winston, New York.

Wright, W. D. and **F. H. G. Pitt**

1935. The colour-vision characteristics of two trichromats. *Proc. Phys. Soc. London* **47,** 205.

Wright, W. D.

1939. *The Perception of Light.* Chemical, New York.
1947. *Researches on Normal and Defective Colour Vision.* Mosby, St. Louis.
1949. *Photometry and the Eye.* Hatton, London.
1958. *The Measurement of Colour,* (2nd ed.) Hilger and Watts, London.
1967. *The Rays Are Not Coloured.* Adam Hilger, London.

Wyszecki, Günther and **W. S. Stiles**

1967. *Color Science: Concepts and Methods, Quantitative Data and Formulas.* Wiley, New York.

INDEX